"McCourt is thoughtful. . . . [He] writes movingly about his family and is open about how his alcoholism devastated them."
—*New York Times*

"Lucky for us, the path of alcoholic self-destruction often makes for a rich anecdotal life, especially when the subject is a glib . . . Irish rogue."
—*Minneapolis Star Tribune*

"*Singing My Him Song* makes it clear that no one understands the value of laughter better than Malachy McCourt."
—*Irish America* magazine

"McCourt offers closure and glimpses of victory over his past and the bottle. . . . But most delightful of all, he recounts his recovery in his own lighthearted way, always seeing the jest in his follies."
—*Miami Herald*

"It's all told with the frankness and honesty for which McCourt has been renowned."
—*New York Daily News*

"McCourt has a wonderful sense of humor and peppers his work with comic, self-effacing moments."
—*Rocky Mountain News*

Nan Melville

About the Author

MALACHY MCCOURT is the father of five
children, and the grandfather of three. He lives
with his wife, Diana, in New York City.

ALSO BY MALACHY MCCOURT

A Monk Swimming

Malachy McCourt

Singing My Him Song

Perennial

An Imprint of HarperCollinsPublishers

First Perennial edition published 2001.

Designed by Elliott Beard

The Library of Congress has catalogued the hardcover edition as follows:

McCourt, Malachy.
    Singing my him song / by Malachy McCourt.
      p.   cm.
    ISBN 0-06-019593-2
    1. McCourt, Malachy, 1931–   . 2. Irish Americans—New York
(State)—New York—Biography. 3. New York, (N.Y.)—Biography. 4. Los
Angeles (Calif.)—Biography. 5. Limerick (Limerick, Ireland)—Biography.
I. Title.
F128.9.I6 M38 2000
974.71043092—dc21   00-059774
[B]

ISBN 0-06-095548-1 (pbk.)

01 02 03 04 05 ❖/RRD 10 9 8 7 6 5 4 3 2 1

I dedicate this book to
Siobhan, Malachy, Nina, Conor, and Cormac, my children.
And to Fiona, Mark, and Adrianna, my grandchildren,
for filling my heart with joy, pride, and breathless love.

A special dedication to some of the warriors of the Willow-brook Wars. It was a place of horror, brutality, and awful suffering with the Stars and Stripes high above on a flagpole in ignorance of the carnage below. When some parents, with the help of some doctors and other workers and a few legal minds, began a desperate revolution, they were called communist Viet-cong terrorists. But that they persisted in the movement proves that all citizens are created equal and have the right to life, liberty, and the pursuit of happiness, even if handicapped, mentally or physically. So, to those who rose up and exposed Willow-brook for the place of despair and death it had become, All Hail! And to the thousands of men, women, and children who suffered and died there, Rest in Peace, and may the atrocities committed there never occur again!

Time, memory, space, and human fallibility prevent me from mentioning all of the Willowbrook warriors and the names of the victims, but here are just a few: Rosalie Amoroso, Eleanore Ash, Dr. Bill Bronston, Kathy Bronston, Bernard Carabello, Tim Casey, Gene Eisner, Bruce Ennis, Ira Fisher, Jerry Gavin, Willie Mae Goodman, Charlie Haney, Connie Haney, Chris Hansen, Jerry Isaacs, Jane Kurtin, Elizabeth Lee, Marie Marcario, Mark Marcario, Anthony Pinto, Ida Rios, Geraldo Rivera, Murray Schneps, Vicki Schneps, Dr. Mike Wilkins, and many more. Thank you for your dedication to the best of humanity, and for your compassion and humor.

Contents

For he comes the human child
To the waters and the wild
With a fairy hand in hand
From a world more full of
Weeping than he can understand
　　—W. B. YEATS

Part I

Behind Bars

On Sunday afternoons in 1963, the summer I worked in a Hamptons hostelry called the Watermill, myself and assorted staff would adjourn to the beach, armed with a largish cooler chock-full of ice, vodka, and orange juice. One of our number, Dan Cohalan, did a creditable job with the guitar, and, as we knew a reasonable number of songs with choruses, we were able to gather quite a number of children around to join in, and their parents were delighted to have us in loco parentis so they could go off walking, swimming, or having affairs in the dunes.

What a joy it was to hear forty or fifty silvery six- and seven-year-old voices raised in bawdy song, and sung with as much conviction as if they knew what they were singing:

> *Oh, I've got a cousin Daniel,*
> *And he's got a cocker spaniel,*
> *If you tickled him in the middle,*
> *He would lift his leg and piddle.*
> *Did you ever see,*
> *Did you ever see,*
> *Such a funny thing before?*
> *D'ye know my Auntie Anna,*
> *And she's got a grand piana,*

Which she rams, aram, arama,
'Til the neighbors say, "God damn her!"

I taught them the occasional limerick, as well.

Rosalina, a pretty young lass,
Had a truly magnificent ass.
Not rounded and pink,
As you possibly think,
It was grey, had long ears, and ate grass.

I can only assume that the parents never asked to hear the new repertoire the silvery-voiced little ones brought home from their sandy Sunday school.

Not a few adults joined us, too, as we were the jolliest gathering on that strand. Two very attractive young women, Louise Arnold and Lynn Epstein, plunked themselves down on the sand at Cohalan's invitation, and soon became regulars. They revealed that they had produced some Off Broadway shows, which sparked my interest. I was of a mind to get serious about the acting trade, due to my newfound penchant for suffering.

It depends on where you are in life, I suppose, but some people think that to be a great actor it's necessary to be entirely miserable, and if misery is the grandest qualification, then it was, *Move over, Burton, Olivier, and Gielgud—McCourt is on the way.*

Sundays were not a joy unalloyed, as every child singing there might suddenly remind me of my own two, who seemed lost to me forever. That summer, my estranged wife, Linda, had informed me that she was going off to Mexico to divorce me. We'd been separated for two years by then, but occasional bouts of blind optimism had led me to believe that it would somehow all work out.

"What about the children?" I had asked her.

"What about them?" she asked. "We never did have anything resembling a marriage, so don't be a hypocrite and pretend we were a family." She spoke truth, but that didn't make me feel any better about it.

One Sunday, when it was too hot to sing, my morbid contemplations were knocked right out of my head, at least for a time. I was enthroned beneath my protective umbrella (this because I have skin which, when exposed to the sun, makes the common beet seem albino), when out of nowhere there hove into my purview the most astonishingly beautiful and graceful woman I'd ever seen in all my life and travels. She had rich brown hair and striking almond-shaped eyes. She wore a modest white bathing suit and, as she stepped along the water's edge on her long lithe legs, the water glinting with sunlight behind her, her slim body and swanlike neck seemed to sway in time to music. Upon her right hip there was perched like a koala bear a bright-faced, blond-haired child in the two-year-old range.

It never occurred to me to think that the presence of a child might imply that there was a husband somewhere; in that moment I was so absolutely smitten that I couldn't conceive that any obstacles might stand between me and this vision.

I don't know how long it took me to realize I wasn't breathing, but a huge exhalation brought me back from near drowning on dry land. Turning to the nearest body on the beach, who happened to be Louise Arnold, I gasped the question, "Who in God's name is that vision walking toward us?"

"That's my friend, Dee, who's visiting me this weekend," sez she.

I was flabbergasted that a mere human being would know this celestial creature.

"I must meet her," said I, "and would you be kind enough to do the introductory honors." Louise was amenable, as she was quite the hand at matchmaking. "Dee!" she called out. "Would you come here for a sec?"

"Dee" came striding over, a somewhat bemused expression on her lovely face. Silenced by the presence of such beauty, I could only extend my own paw to shake her soft hand. The brain and the tongue had disconnected at once, and anything I thought of saying seemed stupid and banal. Finally, I managed to rasp out a "How do you do?" though my tongue felt inert.

Dee sat down and, as I'm not fond of nicknames or diminutives, I ascertained that Diana was her proper name. So Diana she was to me, although old friends and family still call her Dee. I mumbled something about it being a nice day. She agreed. A bit of silence, then I tried again, "A bit too hot, though." She agreed again. An agreeable woman, she was.

Shortly, Diana excused herself, and off down the beach she went, and I was left feeling like a complete ass. "By Christ, McCourt," I said to myself, "for all yer gift of the tongue, for all your much-vaunted charm and gallantry, you couldn't trot out the treasure trove of complimentary clichés you keep on hand in case of being caught without something to say?"

Diana was leaving that afternoon, Louise told me, so I asked Louise for her friend's telephone number, which she obliged me with on the usual matchbook cover. It was a RIverside-9 number, and to this day I haven't forgotten it. I made the vow to ring her immediately upon returning to the city, and I was to carry her number with me everywhere, but I never dared telephone, not even when I was filled with the bravado conferred by whiskey.

This was sacred business, which made my brain whirl and my legs go rickety—and there was still Linda, and the unresolved matter of the divorce. Or unresolved for me, anyway. It was well resolved for Linda; she was divorcing me.

Divorce is an odd thing, but so is marriage for that matter. Although you are generally present at your own wedding there is a certain unreality to the whole goings-on. There you are, having reached the age of reason, standing up in front of two to a couple of hundred friends, relatives, and neighbors, making promises to love, honor, and cherish, to have, to hold, to cleave flesh to flesh, till death do you part. A bloody tall order to ask of a twosome who in most cases have only known each other for a short period. The sort of commitment that if you are asked to make to a member of your own family, or a childhood companion, someone you knew and loved your whole life, your immediate response would be, "Not on your Nellie."

But at least you're physically present at your wedding ceremony, whereas at the divorce, equally momentous, changing your life just as much, it's not always necessary to be at the formal smashing of the union. According to my soon-to-be ex-wife, she was proceeding to Tijuana, our signed agreement in hand, and there a sofa salesman who also repairs mufflers and works as a judge in his spare time will stamp a document declaring our marriage to be over and done—the parties of the first and second parts now free to proceed singly through life without let or hindrance from each other.

It's eerie enough waking up of a morning to be startled by another head on the pillow, who turns out to be your new marriage partner. You find yourself saying, "I'm married. *Jesus Christ, I'm married!* Now what will I do?"

But, then, another morning, you wake up with a vague notion that something important happened yesterday, and it dawns on you, ah, yes, Linda said she was going to divorce me in Mexico. There is no other head on the pillow, but it wasn't just a bad dream, you are unmarried again, but single no more, now you are divorced, such a serious word, even to a recovering Catholic like myself.

Your own mind can be a remarkable thing, telling you things about the world that nobody else standing right alongside you would even suspect to be true. In my years of marriage to Linda, it didn't occur to me that my amorous extracurricular adventures might have consequences. Didn't I come home to her in the end, and shouldn't she be grateful for that? When she'd thrown me out, hadn't I broken into the apartment, ripping the place apart in a rage, and gone to jail for it? Didn't that show I cared? And during the years after, when I'd be off smuggling gold in India, or living on a houseboat on the Seine, couldn't everybody see I loved my children anyway? They were five and four that summer, Siobhan and Malachy, and if I'd been absent much more than I'd been present in the last few years, surely my drunken bouts of self-pity, my maudlin despair at having them taken away from me, counted for something. Why couldn't Linda see any of this? How could she want to divorce me?

I had to avoid that word, "divorce." I was still so bedeviled by remorse over wrecking the marriage I couldn't yet get the sleep I needed without having the drink, and lots of it.

That summer, there was a huge debutante party for a young thing named Fernanda Wood at one of the grander manses dotting the Arcadian landscape. A few of us plebeians popped over after our toils to join the festivities. Apparently, the mini-scions and -scionesses had made a decision to redecorate the Hamptons mansion wherein this bacchanal took place. They had hurled champagne bottles through the windows and attempted to set a number of small fires. The place was aglitter with flitting Caucasians, debs leaping half clad throughout the house, pursued by young bucks garbed in the remnants of tuxedos. Furniture in disarray, glasses smashed, ornaments shattered, chandeliers gyrating to twist music, until some of the higher-spirited lads began to swing from them, bringing them crashing down (predating that famous chandelier scene in *Phantom of the Opera* by many years), while shrieks and screams echoed throughout the house: It was simply the rich at play.

Somehow the press, local and national, got hold of the story, and there were sober articles and tut-tutting editorials re the younger generation and what they were coming to, and rich kids with too much money to spend and too much time to waste. There was a great deal of "In my day, sir" commentary, and headlines containing the words "rampage," "orgy," "volcanic eruption," "riot," "uproar," and every other synonym known to Roget and all the sauruses.

I, being somewhat of an adult, took a different view. I just cheered them on. But it was too rowdy even for us old hands at running amuck. We had gotten to the party just as it was breaking up—in the most literal sense—anyway, and as there didn't seem to be an intact glass to sip from, we departed, shaking our comparatively hoary heads at the wonder of it all.

As I drove down the road, I spied flashing police lights and, not wanting to face judges again, I congratulated myself on making the getaway. I found myself on the Montauk Highway, drunk at 2:00 A.M. It seemed to me a good idea to see how fast I could drive, and off I went.

'Twas a dark night with lacy swirls of sea mist floating toward me as I raced along the empty highway. The only indication of speed was the needle quivering on the dashboard; despite my foot savaging the accelerator, the damn car seemed to be encased in air, immobilized in a Bakelite night. The thought that I could crash and reduce myself to smithereens did float into the head, but I didn't respond, so it left of its own accord.

There were no other cars on the road, no house lights, nothing to tell me I was hurtling to possible destruction. My concentration was on the speedometer and stomping the foot on the accelerator and of course the thought of the soon-to-be ex-wife took over as it generally did in the small hours, spurring me to more teeth-grinding, jaw-clenching, screaming efforts to outrun the demons.

All that tumult being in my head, it took me a while to become aware of the sound of the tires on the road, a sound that seemed to form the words: *Stop it now. Stop it now. Stop it now.* And as I slowly touched the brake, I became aware of the high speed I'd been hitting and suddenly shuddered with the understanding of what this attempted suicide might have done to my children, Siobhan and Malachy, and then I stopped the car.

My friend Steve Epstein had little to do those days so he, as they say, hung out with me. I owed Epstein for letting me stay at his digs back in the city, and there were times in the Watermill when I didn't present a bill to my pal. I would just add his bill to some nonparticular well-to-do type, or forget it altogether.

The summer was moving along slowly toward its end when my tenure at the restaurant suffered a similar fate. One night, Epstein arrived from the city with a girlfriend. I served them dinner and drinks for free, whilst neglecting to make out a tab. The boss, David Eaton, in a fit of sudden efficiency asked to see the tab I was keeping on Epstein. I said, "'Tis in my head."

"Oh yeah," sez Dave, "you write everyone else's down but not Epstein's?"

"Right," sez I, thinking rapidly. "He drinks so little it's easy to remember."

"What about the lobster dinner he had with that broad, and the bottle of wine and the cognacs they had after dinner?"

"All in the noggin," sez I, tapping the side of the head.

"I'm going to have you both arrested," sez Dave. "You for stealing and Epstein for trespassing, as I know he slept on the couch in the accommodations we supplied strictly for your personal use. I'm calling the state police right now, and I know them well."

I slithered over to Epstein and, speaking out of the side of my mouth as I'd seen Humphrey Bogart do in convict films, informed him that if we didn't get our asses on the highway we were likely to be guests of the county.

I was trying to speak in an understandable code yet not give the game away to the man's dinner companion. He was irritated, as he was making great progress with this young thing, and she had already indicated her readiness to have a mutual exploration of the nether regions of their respective bodies, but after I had pulled him into the kitchen and explained the situation the lust left him and fear of being stuck behind immovable bars took over.

What to do with the lust object? Give her money for the taxi and tell her your uncle is at death's door and 'tis necessary to get to New York City to open it for him. I handed over to him my paltry tips to pass along as cab fare, then nipped out the back door and dashed up to the hovel on the roof to pick up the few belongings. I met Epstein in the parking lot, where he was waiting for me as the getaway drivers do in movies. We dropped the young bird off at a gas station where there was a telephone, and then began the flight from justice toward the forgiving arms of New York City. I kept the sharp eye out for the flashing lights of the law while he gunned the engine and got the car up to about ninety-five mph, which was stupid, as it would only draw the police's attention.

It's amazing what the imagination can do when you've had a few drinks and have a voluble tongue to convince the other person of the imminence of a frightful incarceration. I was convinced that the entire police fraternity of Long Island was being mobilized to get us heinous criminals who had cheated a restaurant, and had Epstein believing the same. We were bedeviled, too, by the sight of the gas tank needle doing its delicate dance and gently touching empty, with no gas stations open at that time of the morning. We stopped at every closed gas station and practically sucked out the gas remaining in the hoses. At one of them, we discovered a jerry can half full of blessed petrol, which we stole without so much as a "Sorry to have to do this." I would have stolen it from an old-age pensioner to avoid another night of durance vile. 'Twas that plus faith plus talking nicely to the car, now named Matilda, that got Epstein and myself to civilization and safety.

I deposited the odd bits of clothing in a room in Epstein's digs in Astoria, Queens, New York. The building was owned by the uncle and Epstein's mother, and the lad was living rent-free. There's nothing like a rent-free bed in a reasonably comfortable flat with a roommate who thinks you are the wittiest, wisest Hibernian he has ever encountered, and when I realized that the FBI was not coming after me for an unpaid dinner tab of $19.27 plus tax, I relaxed and circulated once more.

The atmosphere in the apartment was ripeish, to say the least, which I attributed to deficiencies in the housekeeping department, but upon inspection it seemed clean enough for a bachelor's digs. The smell seemed to get worse, though, and finally my nose led me to the epicenter of this horrendous stink. Under the place of my repose, my bed, I discovered a dead crow decomposing and giving nourishment to a full complement of maggots and other guardians of the environment. That foul of the air took its last flight out my window, accompanied by larvae, worms, maggots, and other bosom buddies taking their first and last flight, startling a gossip of elderly women exchanging dark forebodings in front of the building, and giving them fodder for even darker words about the world of dead carrion that flies.

Great barmen, like great hairdressers, are reputed to have what are known as followings; that is, they attract coteries of bods who like the way a bartender talks about sports, or mixes a martini, or in the case of the lasses, the bit of flattery and name recognition. 'Twas said that I had such a following. I wasn't being unduly modest when I said I didn't believe it, as I honestly wondered who in the name of Allah would follow me anywhere. But if Epstein believed this, as apparently he did, and if his uncle was going to finance my reentry into the bar biz, as apparently he would, who was I to say nay, and the search for a suitable premises began.

We found an out-of-the-way spot at 118 East Eighty-eighth Street called the Dublin Bay Café, apparently owned by a Dublin man name Larry Luby. We broached the idea of purchasing the joint but the man did not seem able to say yes or say no. He mumbled something about having to consult with somebody or other, that he didn't really own the place.

The real, but concealed, owners were a couple of shadowy speculators, a husband-and-wife team, smallish stout people named Joey and Tessie who, despite appearances, were brilliant at deal-making in real estate, saloons, diners, jukeboxes, and cigarette machines. Tessie did a wonderful good-kindly-mother act, whereas Joey did a lot of growling and talked about cutting people's balls off and other sporting events of that nature. Because of some legal difficulty with the SLA (that would be the State Liquor Authority, and not the Symbionese Liberation Army, though the one often seems no more reasonable to deal with than the other), they were prohibited from being licensees in any premi wherein liquor was vended.

So we negotiated with them, and settled upon an agreeable price, and all that was needed was the check from Steve Epstein's uncle, which we were assured was a routine matter and would be taken care of as soon as said uncle returned from whatever trip he was off on. But the weeks began to pile up and so did the pressure from the stout people to conclude this deal. I never did figure out whether Epstein was in a fantasy world about raising the money for the purchase of the place all along, or if his uncle just changed his mind in the end. Finally, though, Epstein confessed that his uncle had no intention of financing a saloon for his dopey nephew and his drunken Irish pal.

'Twas left to me to inform the other folk that there was no money in the coffers and the deal was off. They weren't too perturbed, as they immediately had an idea for me. They would shift Mr. Luby to another location, and put me on the Dublin Bay license. I would move into one of the three studio flats in the building. Fine with me, and I immediately took over and changed the name to Himself and did the bit of renovation to make the joint more publike.

While all this was going on, Diana's telephone number remained imprinted on the brain, but I was hesitant to call her, as I was sure she wouldn't remember me. There was many a night I'd take out that matchbook cover and look at the RIverside-9 number, go for the phone, and then find a list of reasons not to call. She wouldn't remember me. She wouldn't be interested in me. There was the vaguest possibility she'd remember me, but suppose she said, "Why are you calling me?" or said she was deeply involved with someone, or demanded—angrily, of course—to know who gave me her number. So, even though my mind was slowly letting go of its obsession with my now ex-wife, Linda, and it came back again and again to a vision of Diana, with all its lovely promise, I never called. I'd say bollox on it and have another evening at the drink as we fixed up the new saloon, Himself.

I'd applied to that wonderfully corrupt agency, the State Liquor Authority, to be the official licensee of Himself and was eagerly awaiting approval, as the only blot on my record was the disorderly conduct charge from having barged into Linda's apartment in a somewhat violent manner a couple of years prior. They'd been handing out licenses fairly freely to the Mafia all over the city, and failing to take them back even if the licensee had someone garroted on the premises or put explosives in the toilet to blow the shit out of an enemy. So, imagine my astonishment when the official envelope arrived to inform me that I was turned down.

It was due to the fact that I had been on another license, five or six years previously, that had been revoked. While still a partner in my first bar, Malachy's, I'd briefly gone into partnership for a minute sum of money with one Lew Futterman in an establishment in Greenwich Village.

Lew, a progressive young fellow, whom I'd met during my rugby-playing career, had noticed that there was no place in all of New York City where couples who were not of the same race could get together to have a beverage and a bite of food without being given bad tables near the kitchen, along with insults and sullen service from waiters and bartenders. He had the logical and commercial idea that were we to open a spot where the miscegenationists could gather, not alone would we be doing God's work, we could make pots of money in the process, because, you see, black folks' money is the precise color of white folks' moola, and has exactly the same value.

I needed a bit of capital for this venture, so I spoke to the missus on the subject, as Linda's parents were well set, and there were indications of a trust fund lurking in some vault. She approached the parents and they, in concert, simultaneously, not to mention together, rose to their full respective heights with an "Aha! We told you he was a fortune hunter, this Mick, and he wants to destroy your fortune while making his."

A bit of tenacity extracted the necessary quids from the claws of the parent trustees, and Futterman and self were in business across the street from the Village Gate, in our new premi, which had a seafaring theme: ropes, lanterns, bits of nets, portholes, and it went under the agnomen "Port of Call."

Having dashed into this venture without careful thought or preparation, and without informing my partners in Malachy's of this new demand on my time, I found myself hoping that I could help run this place without anyone finding out I was connected to it. Futterman had the opposite thought: He was depending on my then celebrity and popularity to draw other folks besides the mixed daters. But before long, my other partners were in a rage because I'd brought disgrace on them with this questionable endeavor, and I was rarely to be seen uptown at Malachy's anymore. I was rarely seen downtown either, for that matter. On the pretext of drumming up business, I was anywhere but in the Port of Call or Malachy's.

The Port of Call was successful, at first, too; there was plenty of money flowing over the bar. Then the screws were applied. The local residents rose in high dudgeon over the dirty goings-on in this saloon, and a complaint campaign was begun. By way of their pressure on the local precinct, we had as many as five visits a night from the police, and there was hardly a night we didn't get a ticket: no soap in the bathroom, no toilet paper in the ladies' room, cigarette butts on the floor, serving minors, insufficient lighting, improper display of license.

Then came the health crowd: cook's head uncovered, a fly on the ceiling, temperature in the fridge too high, meat uncovered, spot of grease on the wall. And the Fire Department: extinguishers not full and in the wrong place, "Exit" sign not bright enough, curtain in bathroom not fire-retardant. We could have had a ticker-tape parade with the tickets we received at the behest of the Mafia chieftains who lived in the vicinity, but Futterman

17

fought on. Myself, I had no stomach for this battle. It scared the shit out of me, and my partners in Malachy's were putting pressure on me to resign, as they said I was endangering our license, so I opted out. Futterman gave me back the initial investment, and we shook hands and parted.

'Twasn't long after that that the mob decided to wage open warfare, and there was a miniriot in the area, the local thugs revolting against the huge black peni being inserted into virginal white vaginas. They smashed the windows of the Port of Call and tried to set fire to the interior, and it was downhill after that. Shortly thereafter, the State Liquor Authority, a perennially corrupt crowd of yahoos, revoked the license, and that was that for the P of C.

When I resigned from the business, I neglected to inform the SLA in writing, so when the Port of Call license was revoked, I was still a licensee, and thus a criminal in the eyes of the Authority. This made me unworthy to be an owner of a saloon in this great and fair city of New York, as I was informed when I applied for my new license.

The real owners of Himself, Joey and Tessie, said we couldn't pull out now, as all our publicity had indicated I was the boss man, so I'd just say I was the owner. My so-called vast following would then trek their way by the thousands, and once more I would be the wise and wealthy lord of all I surveyed, the Malachy of yore.

At the suggestion of one Paul Fagan, another scion, whose family, according to rumor, owned Hawaii and half of the Pacific, I decided to have a formal launch for Himself. There wasn't the extra capital about for such a do, but our cook, Sudia Masoud, a capacious lady of devout Muslim leanings, assembled the sandwiches of cold cuts, and then hordes of black-tied lads and evening-dressed ladies descended on the bar. It was an inelegant joint, not a bit suitable for this gathering of Fagan's society friends, so there was naught to do but get pissed drunk and pretend that it was some kind of joke that called for getting dressed up in evening wear.

Once opened, Himself had the small problem that nobody could find it on an obscure side street of the Upper East Side. And when it comes to running a saloon, the presence of the owner on the premises, whether the real owner or not, is the key to the bit of success. Not far away on Second Avenue, Elaine Kaufman founded the famous Elaine's, still a hangout for the most famous authors and journalists in American letters. To this day, thirty-four or so years since she opened it, there is hardly a night that Elaine is not present to look after her business.

Not so myself. During the renovations of Himself, and after it opened, I was off again, tearing around the city on the usual quest for surcease from the little black demons that used my soul as a venue for their daily outings. The thought of spending the rest of my working life trapped in the confines of a bar was impinging on the consciousness and causing unrest.

When I wasn't running about the city, I'd sit in my monastic room, with the mattress on the floor, the one sheet, one pillow, and the one blanket on the chair, and cogitate on the uselessness and stupidity of it all. Here I was, an intelligent, well-read fellow, curious about the world, good company, easy in society, maybe not handsome, but good-looking enough, with a good sense of humor, with this doomed life's prospect.

During all the time since July, I'd maintained contact with Louise Arnold and her roommate, Lynn. Indeed, while I was still living with Epstein, I was sometimes the overnight squatter at their place when I'd had the snoot-full and didn't feel like facing the trek to Queens. In November, Louise invited me to some party that had to do with promoting a ski resort or skiing fashions, or some such. Having absolutely no interest in skiing or fashions connected thereto, I said, "Of course," and off with me to the party for the free cocktails.

I arrived at the threshold of the large room where the gathering was taking place. A goodly number had assembled by the time I got there. I stepped into that room and 'twas then my life changed forever. Standing by herself against the wall was the beloved Diana of my dreams. She saw me as soon as I saw her and I began to make my way through the thicket of blatherers that stood between us.

At that time 'twas rare for me not to know somebody at these gatherings, but tonight not one soul impeded my path. In short order, I was standing in front of Diana, encouraged by her lovely, warm, welcoming smile. With a greeting I took her hand, mine all atingle at the touch of her, and said, "I will never let you go again."

There was a band playing some music, and I asked her to

dance. With my arms around her, the next words out of my mouth didn't surprise either of us as much as it seems they should have.

"Will you marry me?" I asked her.

Diana smiled and didn't say anything, but she didn't chase me away, either. As usual, I had my ordinary quota of whiskey; after all, a person has to celebrate meeting the love of one's life. (Conversely, of course, a person would have to drink because of losing the love of one's life. Or, indeed, misplacing her, or taking her to dinner, or to bed or to Spain, for that matter, and so goes the nattering, insistent voice of alcoholism.) But I was in good spirits, as they say, and as charming as could be to my newfound, refound love.

I took Diana home that night to what seemed to me a perfect night of lovemaking and was awakened by the gentle touch of her hand on the forehead as she held for me the cup of coffee in her other hand. And there was Nina, the silent, wondering, blond little Nina, just turned two, living in her own world of tongue-clicking and rhythmic head motions, totally baffling the professionals as to her nature. Some said she was retarded; others ventured that she was autistic. Someone else decided on brain-damaged, and one famous specialist said that she was perfectly normal and the only problem was having a nervous mother.

Whatever it was, Nina and myself got along quite well, as I'd grown up with people who bore all kinds of disabilities, physical and mental and emotional (if they are not one and the same thing), and it didn't strike me as anything out of the ordinary.

I took right to this new family, and before long I was quite determined that I would now settle down and take care of Diana and Nina, and try to start seeing more of my own children, Siobhan and Malachy. But I'm the man who gave good intentions a bad name, as simply intending to do something good is no match for derangement and the disease of alcoholism. I still attempted to maintain the fiction that my drinking was harmless fun. I know now that before I could change, before help could be sought and accepted, I had to acknowledge that I had a problem, but I still wasn't quite ready. Love may conquer all, but it does not begin its activities with my timetable in mind.

I'd like to have the proverbial dollar for every broker or banker who has said to me, "I'd like to open up a place like this and have all my friends come and drink there, and I'd get someone to run it for me, and I'd come in on the weekends to say hello to everyone."

HA! I did forbear on most occasions from launching into a blistering response on how hard it is to run a saloon; the syco-phantic air one has to adopt to keep good customers coming back, not to mention the vague unease of constantly selling booze to known alcoholics.

No matter how I looked at it, the reality of what I was doing bashed me in the brain every time. All around the world at dis-tilleries, breweries, and wineries, people were pouring into bottles the fermented results of that which grows generally in fields. Bot-tles and barrels and jugs and jeroboams full of whiskey, beer, wine, vodka, gin, champagne, and bourbon, just to mention a few. And thousands of trucks, trains, ships, and planes were used to transport this stuff to wherever it is needed, and me calling up to order cases of it to sell to my so-called following. Some of them could take it or leave it, but there were others who I knew to be alcoholic (though of course I couldn't see I was one of them), but that didn't stop me from vending the stuff to them. Like any dope pusher, I had fixed expenses, and always, as is said, the rent has to be paid.

There was one lad, Chris, a member of a well-known acting family, who drank enormous quantities of Jack Daniel's every night. He was about twenty-two years of age and was spending on an average one hundred and fifty dollars a week at my place alone. I didn't mind the income, as it literally paid the rent in the early sixties, but I was worried about the damage being done to this young man. So I went to his father, who told me that the son had his own trust fund, and he had no control over it or him. When he found out, Chris was furious that I'd gone to his father. He told me that was the end of his days as a customer at Himself and stomped out.

I ran into him again years later, and he is a sober and mature man, and we are now quite good friends.

Jim Tierney was another one, a tall red-visaged literary type who could quote from *Finnegans Wake*, indeed, from a list of other classics as well. Brilliant as he was, he had a spot of difficulty in holding the job, so he generally hooked up with well-to-do ladies, one of whom was Sally Smith. On the first of the month she breezed in regular as the dawn to find out her beloved's consumption for the previous month and what the bill was. There was great comfort in this until the day she informed me that the free ride had come to an end and no more tabs would be picked up and good luck to you Malachy and off she went.

Jim arrived a few hours later and ordered his usual double Dewar's on the rocks. "A word with you, Jim," sez I.

"By all means," sez he.

"There is a substantial tab due from last month and your lady guarantor has decided not to pay any more of your bills, she tells me," sez I.

"She will get over that," sez he. "I'm delighted we are having this little confabulation," he continued, "as on the way here I was thinking that should I someday own a hostelry such as this, and if you were a habitué, it occurred to me that I should extend to you unlimited credit." He always did speak in flowery terms, which amused me when he was conning other folks, but, as usual, I was in a financial clutch and not in a mood to be so conned myself.

"You don't have a saloon," said I. "And I'm not a habitué, and you owe me seven hundred dollars, which needs paying now."

He slowly shook his head and assumed a disbelieving and disappointed look, and said, "I never thought I would see the day when my friend Malachy McCourt, bon vivant, man of letters, compassionate friend of the needy, would descend to the dungiest depths of sordid commerce by demanding filthy lucre from a man who disdains such transactions. If you persist in your demands I shall have no choice but to leave these premises and, when I do, I assure you I shall never grace your porte cochere again!"

Jim stalked to the door, opened it, turned dramatically, and, in stentorian voice, he bellowed, "And furthermore, *FUCK YOU!*" He marched off into the night, leaving me with the gob agape and somehow feeling that I was guilty of something.

That other Limerick git, Richard Harris, had just finished playing King Arthur in the movie version of *Camelot*. Harris had grown up among the toffs in Limerick, not part of my crowd, but I'd encountered him there once, in a game of rugby, and he had been among the Irish and British actors that made my first bar, Malachy's, home base. In a fit of noblesse oblige he now decided to once again move among us common people.

Wearily he told me he'd had his fill of the chicanery and falsity of Hollywood and the acting profession and that what he would like to do is work for me as a bartender. So it came to pass the man got behind the stick with another stalwart man, Jack Sandon, a fine barkeep, who never removed the cigarette from the corner of his mouth even to deliver the most stinging of insults.

Harris poured with abandon and without measure and never seemed to take money from any of the clientele, and there were many more than usual, as word got out and the dazzled came to gaze at this movie star boniface. One eve, a couple of cheery and quite inebriated elderly ladies told me that my bartender was a very nice young man as he refused payment for the bottle of Dom Perignon they had imbibed. The hand was clapped to the forehead on receiving that piece of news.

But at the end of the week the King was surfeited with serving hoi polloi and gave me notice that he was quitting and going back to London. None too soon, sez I to myself. But there had been a deal. Harris had instructed Jack, my other barman, to write down all that he gave away, and at the end of the week he gave a check to Jack to give to me after he had left and that check covered all of what I had thought were free drinks. A generous man.

I continued my brooding through all of it, and couldn't see any way out of the dilemma of making the living. Running a public house (from which the word pub arises) means you are open to the public, and have to be prepared to greet any and all who walk through the portals, be they drunks, arseholes, fools, convicts, prostitutes, Wall Streeters, laborers, cadgers, the sad, the bad, the glad, and, horror of horrors, the boring. One of these had gotten my ear one quiet night, and was describing his work as a salesman of steel products and proudly showed me his business card, which was made from rolled steel. That was it, too much for me, so I hied my way to the back room to get away from the rigor mortis of his talk.

I was alone in the back room a little while later, dozing at the table, when I heard a commotion and a voice shouting in the bar, followed by a gunshot. A small parade entered the back room, where I had placed myself under a light so as not to startle the gunman. The little procession consisted of Jack Sandon, barman Ally Cobert, a serious waiter, the hilarious Bob Boland, who was also a waiter, a customer lady, some unknown man, and a young scion named Thomas Fortune Ryan, all with arms and hands well up into the air. They were followed by two lads of African-American descent carrying guns, who made loud and frequent reference to the fact that all of those in the assembled group had had some kind of sexual relations with their mothers.

We were seated at the various tables and told to keep our hands in sight. One of the gunslingers sat guard whilst his pal went out front to get at the cash register. The conversation did not touch on anything of importance, no reference to current affairs, theatre, or literature. Indeed, it was more demanding. Orders, in fact, emanating from our hold-up man. To wit, "Empty your motherfucking pockets," to the men, and "Gimme that purse, bitch," to the only woman in the group.

The cash register ransacker came back swearing that he couldn't open the motherfucking thing and some motherfucker better come and open it or some motherfucker was going to have his motherfucking head blown off. Jack offered to do the job for him and, while they were out of the room, Fortune Ryan asked our captor if it were permissible to smoke.

"Go ahead," sez the gunman.

Fortune R. picked the cigs out of his shirt pocket and, being a well-bred lad, offered one to the armed friend, who seemed highly offended that anyone would think he was a smoker. Ryan apologized for his assumption and timidly asked if the man had a match. The lad raised the pistol and told Ryan to put the cigarette in his mouth and it would be lit with a bullet.

Jack returned, along with the other fellow, who complained about the paucity of money in the register, and then shouted at Bob Boland to stop looking at him, and fired a shot past his head into a mirror to emphasize his point. We all looked pointedly elsewhere. Then we gallant six were herded into the cellar as the duo announced they were going to work Jack over until he revealed where the rest of the money was concealed. I said there was no more money, and I should know, as I ran the joint. The cold rim of a gun was placed at the right side of my head an inch from my eye and it was pressed hard into the skin.

I couldn't help but think that a simple pressure on the trigger was the next step, and that through that cool barrel would travel a sheathed bullet at a great blasting speed, entering my head, tearing and rending the flesh, the bone, and the brains, scattering them and splattering them on floor, walls, and ceiling.

Closing my eyes, I said good-bye to Diana and Siobhan, Nina, and Malachy, but my good-byes were interrupted by a snarling voice ordering me down the stairs, still alive, to my astonishment. I'd always wondered what I'd do when faced with the possibility of immediate death. Some people say you'd pray, beg for forgiveness, beg for your life, plead with God to save you, but I found myself strangely without fear, as if this were happening to someone else, a trifle curious to know what it is to be shot and to die.

In the end, the bandits didn't beat Jack, as they finally got the idea that there was no more dough, and off they fled into the night. Ally Cobert promptly locked the front door after them, which led Jack to ask him if he'd ever worked in a stable. Then, when the police arrived and asked how much money had been stolen, Bob Boland told them not much, but I'd written them a check for the rest.

Despite the fist-sized cloud of Vietnam, hanging low and menacing on the horizon, the sixties had come up smiling, with JFK and the charming Jacqueline riding waves of adulation from cheering and cheery crowds everywhere. Yes, we'd had the Bay of Pigs, but that was wriggled out of and had been planned by the Eisenhower crowd, egged on by Nixon. There were some nasty confrontations in the segregated South, but President Kennedy kept the lid on that boiling pot, and on Khrushchev, and on anything else unpleasant brewing in the world or beyond, as the Mayoman said.

I was in my monastic bed in the apartment above Himself when the phone rang around noon on November 22, 1963. 'Twas the soft-spoken Diana asking if I'd heard any news on the radio about the president being shot. My tendency is always to move into comforting mode, so I said it was probably a mistake and that there would be clarification very soon.

It wasn't a mistake, and the clarification came much too soon. The man had been shot and he was dead. Within me, I had held a pride that an Irishman had made it to the White House, and it told me that America was opening up to me, too. There was a wit about the man, and the way he would poke fun at himself and the brothers made me think he was like me, someone I could have a drink with. When he was shot, it felt as if it had been done also to me, as if they had told me that the dreams I had for the future and my life in America weren't possible.

If you could collect a dollar for every time the words "I can't believe it" were uttered in those gloomy days, you would be among the wealthiest of the world's denizens. We, Diana and myself, spent all that weekend together cementing our love in the grief of the day. We walked, talked, played with Nina, turned the television on and off and on again. Listened to people raging on the radio as to whether ball games should be canceled, whether Broadway plays should stop, was it profoundly disrespectful to go to movies. I did manage to get to work, but Himself was empty, the gloomiest place to be. I went to P. J. Clarke's at one point, to immerse myself in a crowd, but it was nearly empty, too. There was a lot of staring into glasses going on during that weekend. I'd look up, shake the head in disbelief, say something inane, and go back to staring. One fellow stood up in the back room at P. J. Clarke's and announced that if anybody said anything against President Kennedy he would deal with them personally. Needless to say, there were no takers. Of course, there were mutterings all over town about conspiracies and dirty doings by Nixon, who had been in Dallas that morning, and about Johnson and the coincidence of the assassination taking place in Texas, his home state. Then came the arrest and killing of Lee Harvey Oswald, leading to a confusion that has never been dispelled.

But time passed, as it always will, and everything eventually went back to normal, or whatever passed for normal. Diana still smiled and remained silent when I'd bring up the subject of marriage. She was virtually a prisoner at home, having to take care of Nina, and was still trying to get a straight diagnosis on whether the child was retarded, brain-damaged, or autistic, and she was still not getting one.

Diana was, and remains, the most remarkable woman I've ever met. As a young girl, she had studied ballet with George Balanchine, at the School of American Ballet, and attended the

Professional Children's School in New York City, whose curriculum was designed for kids involved in show business. Her father and mother, John and Bernice Huchthausen, encouraged the odd schooling despite the long commute from Ossining, in Westchester County. Diana got a scholarship to Smith College, from which she eloped shortly before graduation. She went into the publishing business, as a foreign rights manager at Harper and Row, and started up a literary magazine with her husband. But then came Nina, and then divorce, and she was now limited to taking in typing, which was somewhat akin to taking in washing. She wasn't even that good at it, and didn't really care to be, but she did type *Catch-22* for Joe Heller. He paid her as an act of charity, she sez, as her work was quite bad.

We spent many nights together, but there is no denying that on drinking nights, when the opportunity presented itself, fidelity, never my strong suit, was right out the window, without a second thought. Whiskey was and is a wonder to me in that it made me comfortable enough to be something of a lady's man, and it transformed me in my mind from a guttersnipe to a wit, a sophisticated, erudite man-about-town. I prided myself on never stuttering, stammering, or stumbling in the course of an evening's peregrinations. I had the ability to speak the most arrant nonsense and appear as if I were in command of facts and statistics to confound any listener.

There was a night when I did a long monologue on the accomplishments of Leonardo da Vinci, ending it with a peroration on the magnificence and beauty of his sculpture the *Pietá*. Some know-it-all spoilsport piped up that it was Michelangelo had done the job. I tried to oil out of that one by saying that I wanted to make sure everyone was paying attention.

Late in 1964 Diana suddenly told me, quite upset, that she didn't think our relationship was going anywhere and that it had to come to a halt. She had to look out for herself, she said, and it was true that I was taking her very much for granted. I gave her no sense of commitment, but assumed that she would always be there whenever I was ready to grace her with my company. Not infrequently, I didn't bother to show up when I said I would. Nonetheless, this was completely unexpected, and I was stunned. Not having a terrific speech ready, I agreed we should separate.

There followed days of grief, anger, and sorrow over my latest loss, which of course called for some serious drinking. When I thought about what Diana had said, in my few sober moments, I had to agree she was right to be quit of me. Here I was, stuck running a smelly saloon that not only was losing money, but was a totally illegal operation anyway, as the man on the license was only a front. We were always late with our taxes and with Con Edison, always failing health inspections because a damn sewer pipe was leaking into the cellar, where large gray rats didn't bother to scuttle off when we came down for beer and supplies.

Sometimes I'd have no money left to pay myself after the secret owners came and took their weekly share. I was trapped in this place by my fear and self-loathing, feeling savagely inferior to everyone around me. There didn't seem to be any exit in sight.

Now, the woman of my enveloping dreams, the woman who seemed to hold out some hope of a future, had seen fit to leave me because our relationship was going nowhere. I managed at frequent intervals to curse God and the donkey he rode in on.

But for once in my life, instead of saying, "Bollox on it!" I took a positive action. After a week of this, I picked up the telephone and called Diana and poured out from my soul a torrent of love, of loneliness, of longing to see her and be with her again. I yowled that I would lay down my life for her, that all I had was hers and that she must marry me.

There was a silence on the other end of the phone, and then that gentle voice spoke, saying she had missed me too. "Yes," she said. She would marry me.

"When, when, when," I said, rushing headlong.

"December first," she said, after a moment's thought.

It was September then, and as soon as I realized how little time there was between then and now, I slammed on the brakes. "That's too soon," said I. From the loneliest man in the world to the most terrified: elapsed time, two seconds.

"All right then, when would you like to get married?"

"March first," I blurted, for no good reason.

"That's fine," sez my beloved, and so we were engaged and committed to say the I dos and live happily ever after.

Ha.

Of the bad habits available, I missed very few. I drank too much, ate too much, philandered too much. I had managed, though, to somehow remain a nonsmoker, a state I remedied at about that time. There were still commercials for cigarettes on television then, and an advertising campaign for Lark cigarettes featured a truck traveling around the country with someone on board shouting, "Show us your Lark!" to people in various walks of life.

I auditioned to be one of the sham workers and, not being a smoker, I had to practice. I reasoned that I'd never get addicted like my mother and father before me, as I really disliked the damn things, but in the course of doing the commercial I got hooked. I got paid around three hundred dollars for the day's work and proceeded to spend thousands of dollars to maintain my new habit, not to mention my damaged health and yellowed teeth and the hundreds of little burn holes I put in various garments (my own and others') over the years.

I also got to do some other commercials during this period. I played Henry VIII for Imperial margarine and again for Reese's peanut butter cups. Large, bearded Irishmen seemed interchangeable with English kings on Madison Avenue. My pal, Dick Hope, husband of the witty Marilyn, took up a professional challenge one night at the bar, to wit: Could he create a commercial for his client's product, Colgate-Palmolive lime shave, using me, a bearded man. Not only did he do it, I got the part. What he had me do was act the bartender role (less a stretch than Henry VIII) and squeeze a lime into a drink. Instead of lime juice, out comes shaving cream, which I lathered onto my beard, saying, "Now why would they go and tempt me to shave?" A poet, a scholar, and, above all, a decent man was Richard Hope.

I also found myself a panelist on *The David Susskind Show*, a syndicated television program that had a huge viewing audience. This particular show had as a theme folks who had to deal with the public and the difficulties they encountered. There was a waitress, a hairdresser, a taxi driver, and myself, from the saloon business. As was my wont, I had fortified myself against vocal aridity with a few jorums of whiskey.

Susskind was his usual expansive self, very sincere, trying to accommodate the nervousness of the neophyte panelists. Many successful people get the backlash from the begrudgers, and David Susskind did not escape. In those days, people were quite vociferous in their opinions of him, which were quite low, similar to those who speak ill of Geraldo Rivera in this generation, saying he's not to be taken seriously. However, it was not generally known that this man Susskind, a successful producer of television shows, movies, and Broadway plays, employed many of the writers and performers who had been blacklisted by the Hollywood and congressional scumbags, and risked his own career in doing so. I believe he should be judged by the good he did, which was quite a bit, and more than enough for me.

On this panel, the talk wandered about the table—complaints about the vagaries of the public, and the stupidity of certain segments thereof, the paucity of tips, and the insecurity of jobs. There were calls from the public as well, one of which was from a hairdresser who could only be described as extremely effete in manner. He complained that because of his profession, he was always being teased about being a homosexual (the word "gay" still being public property at that time), though he said he wasn't. He added that he had ample proof of his manhood, being an ex-Marine.

The gruff New York taxi driver who sat beside me said, "Why dontcha wear your Marine uniform while you're woiking?" The image struck me, in my somewhat liquored state, as so funny that I began to laugh and couldn't seem to stop. As I leaned back in my chair, it broke, tumbling me to the floor, helpless, on national television, with the cameras following me. Eventually, I recovered, got back onto a new chair, and continued the discussion.

What I didn't know was that Diana had alerted her mother and father, who had yet to meet me, to the fact that I was going to be on the show. Her father's response the next day was, "You are going to marry *that*?"

Diana's parents, John and Bernice Huchthausen, didn't exhibit a wholehearted acceptance of me at first, and understandably so. That had been their first glimpse of me, drunk and falling off a chair on national television. Not long after, Diana and I spent a night together at the parents' apartment while they were safely away in the country. We thought. Early the next morning, sounds of a key being inserted in the lock heralded the arrival of the mother, who was quite shocked to see her daughter in the parental bed in the company of a naked, bearded man. There was a grim set to the lady's jaw and a steely glint in the eye, which I felt boded ill for our future relationship.

For all that, though, things did get smoothed out. I wrote a letter to Bernice apologizing for the seeming insensitivity and tawdriness of the in flagrante moment and vowing the honor of my intentions. She seemed to accept the apology.

I liked Diana's parents, and her sister, Heidi. Diana's father, John, an architect by profession, was also an amazing classical pianist. He wrote music, painted, drew cartoons, wrote poetry, and designed Christmas cards. He was very whimsical on occasion, too, a trait not usually associated with folks of German origin. He was one of ten children of a Lutheran minister from Minneapolis, but he wasn't at all hidebound by religion or by convention. He remained to the end of his tenure on earth a New Deal Democrat, and there was no saying anything against FDR.

Bernice, his wife, was of Swedish origins and working-class background. Her family name was Engstrom. She had studied art, interior design, and architecture, but, as a woman, she encountered restrictions in entering that last profession, and became an interior designer. Still, not bad for the children of Swedish and German immigrants.

After those initial, bumpy, encounters, we all got on fine. I never told or countenanced any mother-in-law jokes, either.

The situation in French Indochina, or Vietnam, as it properly came to be called, was looming ever larger on the horizon. Lyndon Johnson decided that an errant floating log was a torpedo that had been fired at a U.S. destroyer, and persuaded Congress to grant him power to carry out any military action he wished under the Gulf of Tonkin Resolution.

I'd read a bit about Ho Chi Minh and his struggle against the savagery of the French colonials, and I knew he'd assisted in the war against Japan, so I was shocked to learn the U.S.A. was now attacking this patriot. Charles E. Martin, a cartoonist for *The New Yorker*, and his wife got me involved in my first antiwar demonstration in 1964. People on the sidewalks screamed at us and threw things, calling us scum, traitors, commies, and perverts, and letting us know that if we didn't like it here we were strongly urged to go to Russia.

I didn't know enough about the issues to really debate them, but I did know that the Vietnamese people had a right to live in their own country, and the French had that same right, only in France. Looking at those faces, twisted with hate, I wanted to tell them that it was their sons who were the likely dead and wounded victims of this war, and that they should join us to help stop the inevitable mass murder.

Little did any of us know that it would be more than a decade and three presidents later before it was all over. There would be fifty-eight thousand U.S. dead and a quarter million wounded, and several million Vietnamese dead and maimed before a semblance of peace would be restored.

My friend Hugh Magill and his wife had arranged for a justice of the peace to marry Diana and me, on Monday, March 1, 1965. Louise Arnold, who had introduced us, now married to John Westergaard, a lovable, eccentric bear of a man, joined us for the mini-ceremony, as did Diana's mother and father.

We have only one picture of the wedding, taken before we left for the house of the justice of the peace, a man who bore the unforgettable name of Euclid Shook. I think he and his missus must probably have had a martini or two that evening, as they were an unusually jolly couple, offering around the beverages, as we were in their home.

After the I dos, Diana, now McCourt, and self sped off to some old inn in Hartford, the Old Forge, I believe it was called. For two people who had both been married before, we were a shy couple that night. We turned on the television for comfort and diversion, and there was a movie playing which I fervently hoped would not portend our future. It was *I'll Cry Tomorrow*, with Susan Hayward, as dreary a film as you'd ever see and hope to miss.

In the morning I managed to get the car stuck in a snow bank, from which we were rescued by a French Canadian couple. Another stop, just a little later, to get in the backseat and steam up the windows, and then back we went to reality and life in New York.

At that time there was no housing crunch in New York. Newly built apartments were plentiful on the East Side, and the older and bigger apartments were available quite reasonably on the West Side. We opted for one on the West Side, with the several bedrooms and, as they say, two and a half baths, and they were just as glad to get us as tenants then as they would be glad to get rid of us today, as we are still there, and they could double or triple the rent as soon as we left.

We were both moving from relatively small places, and this new habitation seemed huge and full of echoes. We thought we would never be able to afford to furnish it. But Diana had some furniture, and I had access to a knife and spoon and a few things like that, so we set up housekeeping with what we could.

Merv Griffin had started his syndicated television show, with Arthur Treacher sniffing superciliously at all the vulgar goings-on while offering the occasional witticism (he told me that, secretly, he was having a jolly good time). My friend Tom O'Malley, possibly the best talent booker in the business, was involved from the start, and so I had a reasonably good run as an irregular regular with the show.

There is the illusion that all these chat shows consist of spontaneous and impromptu conversations between celebrities who know each other very well. Not so, old sport! All guests, no matter how well known, are prepped, as they say, by a talent booker. Particularly young actors and actresses ill read and lacking in wit, which is more often the case than you'd want to know. Vaguely humorous anecdotes have to be drawn out of them and inflated into stories, and then polished by the show's writers until they are actually funny, or else the whole interview is apt to reveal how boring the guests really are.

I, of course, was the ideal guest, replete with the story, the jest, the bon mot, or so it seemed to me. Griffin liked to come to Himself after the show, and there were nights there with Dom DeLuise, Jonathan Winters, Pat McCormick, and Jack Burns that can neither be remembered nor forgotten.

In the kitchen, the cook, the big-bodied, laughing Sudia Masoud, my favorite Black Muslim, eavesdropped all night and added her shrieks of merriment to the general uproar. She had been present when Malcolm X was shot down, and told me, "That was the cleanest assassination I ever did see." I forbore asking her how many others she had witnessed.

Diana developed a vague suspicion that she was pregnant, and a visit to the physician made it a certainty. We were told that a new child would make its way into this world sometime around the middle of October 1965. I informed my mother, Angela, that she was about to become a grandma again, and she launched immediately into the keening mode.

Now, for those who don't know, keening is an ancient Celtic expression of grief or sorrow, usually heard at a time of death. It is expressed by a high-pitched wailing sound with mourners beating breasts and giving vent to the odd shriek in the middle of the wail. While it was not quite the full frontal keen, the mother did a fairly good job moaning about what would happen to the other children, Siobhan and Malachy. If I couldn't look after them, how was I going to look after the new one?

There is nothing more aggravating than someone giving voice to your own unspoken fears.

We weren't doing well financially, and we were trying to cope with raising a handicapped child. Plus, I'd made a *haimes* of my role as father to Siobhan and Malachy, so I had my own doubts. I contributed what I could, but Linda took care of our children largely with money she got from her parents, who had quite a bit of it. Having settled in with Diana, I saw Siobhan and Malachy, now six and five, most weekends, but I was as apt to bring them home and then go out, leaving their care to Diana, as I was to stay and give them any of what they needed from their father. During their earliest years, I had been completely absent a good amount of the time, sometimes just too drunk to show up.

But the mother Angela was never comfortable with the women any of the sons married anyway, and announcements of pregnancies only served to deepen her gloom that liaisons were going to be on the permanent side. Yet when babies shouldered their way into the world, the mother became most maternal and loving, at least until the little ones reached the age of two or thereabouts. At that point, they got a bit of independence and she'd shift her attention onto the next infant.

On the evening of the thirteenth of October 1965, Diana announced that there were certain movements within her body that indicated a desire on the part of someone to take his leave of the womb. So it was off to the New York Hospital with us. We had taken some Lamaze classes with a lady named Elizabeth Bing, the natural childbirth guru, and for the first time, I had the sublime experience of watching the new life make its entry into our orbit. It was a boy, whom we named Conor Turlough. He was a long lithe fellow, and of course the most brilliant baby in the nursery.

Nina, my stepdaughter, wasn't making much progress, and the experts were now saying she was very retarded. She sat for long periods of time, crinkling cellophane paper from cigarette packs and rocking back and forth. She was for some reason terrified of solid foods, so even when she was six we were still getting her jars of baby food and spooning it into her mouth. It occurred to me that as she had all her teeth and seemed otherwise in good physical health, perhaps some solid grub might be in order.

I apprised Diana of my intention and suggested she absent herself and Conor from the house, as I didn't think it was going to be quiet or pleasant getting Nina to eat the hamburgers I'd prepared. (Were I doing it today, I'd probably select rice and beans or tofu, as I'm a vegetarian, on health grounds.) I sat Nina on a chair at the kitchen table, tied a large apron round her neck, and spread out some newspapers on the table and on the floor, and so began the battle.

I'd made about eight medium-size burgers, which I broke into bite-size pieces. I popped the first piece into Nina's mouth, where she allowed it to rest for a brief moment. When she realized what I'd done, her eyes opened up wide with fury and rage at this big person who'd forced foreign matter into her mouth. She let go with a yowl and spat out the offending morsel, which landed on my shirtfront, leaving a stain before descending to the floor. She quieted down, and I tried again, putting another piece of hamburger in her mouth, all the time speaking as softly and as soothingly as I could. Same result: Out came the meaty projectile, which just dropped to the floor. We sat for a while, me doing all the talking, as Nina did not and does not have speech.

Nina made no attempt to get off her chair, nor did she keep her mouth shut tight to prevent me popping in the food. There were times during this hour-long battle that I was sure I was being bamboozled by this child, as she yowled without conviction, and her resistance was confined to spitting out the food. She'd sometimes have a look of disdain and amusement at this hulk of a man trying to feed her. Somewhere I had read that Annie Sullivan, who took on the task of teaching Helen

Keller the rudiments of ordinary societal behavior, had had a similar siege and, heartened by that thought, I continued the routine. Pop, spit, talk. Pop, spit, talk. Pop, spit, talk. The kitchen floor and table were littered with hamburger. Splatter after splatter, it appeared soon enough as if it were raining hamburger meat in the kitchen.

I was about to admit defeat and sue for terms of surrender when my doughty and noble opponent decided to have mercy on me. She retained a chunk of burger behind closed lips and smiled her Mona Lisa smile at me, still not swallowing, but after a long, long interval I noticed little movements that indicated something was headed toward the stomach, and that's how Nina ate solid food for the first time in her six years of life.

But this small step forward with Nina was just that and no more. With the new baby, Diana was overwhelmed, and there wasn't anywhere we could turn for help. There were no day programs suitable or, indeed, willing to take Nina. When we tried to take her out, she would stage screaming sit-down strikes on the sidewalk. We began to think about permanent residential care.

It is not an easy or simple decision to admit that you cannot raise your own child, but in the end, that is what we did. We found a small home in New Jersey run by a very kindly, bright lady and, soliciting all the financial help we could from family, we arranged for Nina to live there.

It was a bright, sunny day driving out there, but it was hard to appreciate as Diana was teary-eyed and heart-sore at the prospect of parting with Nina. It was hard for me, too, as I'd become very attached to this sweet, trusting child. Nina played quietly in the backseat, not knowing she was heading for a totally new life. When we dropped her off, we saw that a couple of the other kids were similar to Nina in age, condition, and behavior, something we took a bit of comfort from, because we figured Nina would not be an unknown quantity. In a sense we were now free—free to be married, to travel about, to go out, to be parents to Conor—but the price was high. As we drove away, we stopped to look back and saw Nina with her new mentor, standing on a rise outside the house, the sun lighting up her face and turning her blond hair to a light, golden aura. We both wept, because no matter how often we visited her, we knew that child would never live in our home again.

Diana's sister, Heidi, had married her high school love, Warren Washburn, a Marine, and they had become parents to Kelly, a brilliant little girl. Being married and a parent was no barrier to service in Vietnam, though. Warren, a charming, gregarious, devil-may-care sort of lad, assured Heidi and his family that there was nothing to worry about, a statement which can be depended upon to cause a lot of worry.

I don't know why I felt so strongly about that war, particularly as I didn't have to go there and slog it out myself, but I did. The French had treated the people abominably, with the usual colonial torture and murder, and then pulled out, leaving the U.S. to carry on the savagery. Of course, being steeped in Irish history and the brutal centuries-long occupation of Ireland had an influence on me. Colonial powers were always brutal, I knew, from what I'd seen and heard growing up, and what the French, and later the U.S., did to the Vietnamese seemed to me little different from what the English had done in Ireland. Indeed, I could never understand Irish people who supported that barbarous and diabolical attempt to bomb poor people into accepting an alien culture.

Whenever Richard Harris hove into town, I got the inevitable visit at the saloon, and, of course, we went on the inevitable bender. Nearly always there was a brawl. Harris always drew attention wherever he went, and it was astounding to me how many "tough guys" felt the need to challenge the man to a fight. He usually charmed them out of it, but there were some who wouldn't be charmed.

One night we were carousing in some club, and perhaps we were too loud, and perhaps our language may not have been of a pristine quality, but it was directed at no other humans but ourselves. So, when a blocky sort of fellow stepped up to our table and ordered us to "shut da fuck up," it was an astonishing interruption to our fun. I believe we both said, "*What?*"

"You hoid me," sez the blocky fellow. "My lady don't like dat kinda talk so shut da fuck up."

I ventured the opinion that if she was a lady she wouldn't be listening, and he glared down at us from across the table. "Fuck you, buddy," sez he to me.

Without so much as an "After you, Claude," Harris and myself dived for this belligerent lad, overturning the table. We got ourselves tangled in the tablecloth, swinging, punching, and swearing, only to discover, upon becoming untangled, that our provocateur had departed the premi with his lady, escorted by the owner and barman. Celebrity had once more won out over civility and good manners.

Some years before, when I was visiting Harris in Hollywood, I had gotten arrested for trespassing when I went for a midnight swim in his pool. He'd left me sitting in jail overnight, which he thought was very funny. I had not been quite so amused, and even though years had passed, I thought about revenge from time to time.

One day, during that same sojourn in New York, Mr. Richard Harris rang me and invited me to lunch at the Sherry-Netherland Hotel. When I arrived, I noticed outside a small throng of people carrying notepads and cameras. They were a minor mob of autograph seekers who'd heard Harris was in residence.

I proceeded to the Harris suite and was greeted warmly by the man himself. He got room service to bring up the lunch, and we were chatting leisurely, when the star suddenly bounded to his feet and bellowed that he had just remembered an appointment and must get in the shower, and would I answer the phone if it rang?

Moments later, Harris already showering, the phone rang and the concierge asked to speak to Mr. Harris.

"Speaking," I said.

"The fans here want to know if they could send a delegation of three up to your room to get signatures and pictures for all of them."

"Three?" I asked.

"Yes sir," sez he.

"Send them all up."

"But there's more than twenty of them," he protested.

"That's okay," sez I.

"It's your funeral, Mr. Harris," sez he.

In the bathroom, the star-turned-singer had launched into

the song "How to Handle a Woman," from *Camelot*. When the bell rang and I opened the door, Harris warbling in the background, I was faced with about an acre of wide eyes and acne. I invited them all in and told them to make themselves comfortable and to help themselves to anything at the bar. I asked them to tell Mr. Harris that Mr. McCourt was called away on some urgent business and then I left.

When Harris regained his sense of humor and resumed speaking to me, he told me that he'd walked out of the bathroom bollox naked to discover that a considerable number of the citizens of New York were comfortably ensconced in his hotel room. He was not too amused by my prank, but I explained to him 'twas a small price to pay for having left me in jail overnight in California.

'**T**was about then that I was temporarily rescued from the suffocating torpor of the saloon. One day, a very well-dressed man approached me at Himself and said he would like to talk to me about hosting a television show. There was a need, he said, for a natural kind of chat show with some humor and bite to it. I admit to having peered about to make sure he was talking to me, and indeed he was. I said I'd be delighted to do it and hands were shook and commitments made.

As it turned out, the first show had to be taped in California, because by the time I was scheduled to go on the air, I had fortuitously become involved in a movie as well. Richard Harris had introduced me to Marty Ritt and Walter Bernstein, a couple of stalwart survivors of the blacklist in the McCarthy years, who were now making a movie called *The Molly Maguires*.

The Molly Maguires were a group of Irish coal miners in Pennsylvania in the 1860s. They suffered the usual economic terrorism at the hands of the mine owners, and when they attempted to organize, they were brutalized and fired. 'Twas said they had resorted to the bit of sabotage, and knocking off an occasional payroll, and for this some of them were arrested and hanged.

Sean Connery played the leader of the Molly Maguires (the name, by the way, lifted from a secret society in Ireland who were said to dress as women so as to escape detection when they made their forays against the Brit landlords), a man by the name of Black Jack Kehoe. Richard Harris played James McPartlan, an Irishman and an informer, who was paid to infiltrate the group and report on their doings.

Paramount Pictures took over a little village called Eckley in Pennsylvania, where little was required to take it back to looking as it looked in 1860 except to bury some telephone wires and spread some coal dust on the dirt road. On either side of this rough little road were still standing old company houses, little cabins, each with a garden patch complete with the outside lavatory in back. They had small porches out front, and on good days 'twas not unusual to see a retired miner sitting there coughing what was left of his life away, now that the coal dust had done the dirty work of destroying his lungs.

Pneumoconiosis is the doctors' name for this incurable disease, otherwise known as black lung, or silicosis. I was fascinated by these men, who spoke proudly of being miners, never regretting working in the nether regions, but dying all the while. They couldn't get comfortable, these old men, either sitting up or lying down, and their breathing was so tortured it caused palpitations in my own chest from my empathetic efforts to help their respiration.

Just as surely as those miners' lungs had been destroyed by their years in the mines, parts of that beautiful country had been destroyed by strip mining. Thousands of tons of coal dust had been spread over thousands of acres of lovely green land, giving stretches of it an appearance not that different from the surface of the moon. Though it was an abomination to the environment, it gave Marty Ritt a wonderful opportunity to shoot an oddly colorful but dead and deadly landscape.

So, to this forgotten town came a film company with lights, generators, rain-making equipment, cameras, trailers, a hundred or so crew members, and the actors Connery, Harris, Brendan Dillon, the beautiful Samantha Eggar, Bethel Leslie, Frances Heflin, the terrifically talented Anthony Zerbe, the huge Art Lund, and a grand actor named Frank Finlay who came over from Britain to play the part of the evil police chief. I had been cast as—what a surprise!—the local saloon keeper, in whose premises the Molly Maguires plotted their dastardly doings. My part had me attending wakes and going to mass to hear the priest condemn secret societies like the Mollies because, as usual, the Church was in cahoots with the capitalistic savages who were murdering the miners.

I also had to referee a strange football-like game invented for this film, a combination of soccer, rugby, basketball, boxing, wrestling, and mayhem. We played that scene on a blistering hot July day, on a field bereft of grass and rock hard. We were all dressed in the heavy woolens of the period, resulting in people falling down from heat stroke and exhaustion. To end the game, I was instructed by the director to run to a certain spot, look down the field, look at my pocket watch, blow my whistle, and wave my arms. I was so confused with heat and dust that, when I got to the designated spot, I took out the whistle, looked at it, and then attempted to blow the watch, which caused a huge hoot of laughter amongst the hundreds of spectators and players.

It was a most convivial company, with most of the shooting taking place during the day, and dining out and storytelling in the evenings. Marty Ritt had his story of being blacklisted and anecdotes brought back from Hollywood. Walter Bernstein had been in the army during the Second World War, on the staff of *The Stars and Stripes*, and had slogged his way through German-occupied Yugoslavia to get an interview with one Josip Broz, a.k.a. Marshal Tito. Of course, he had his own blacklist stories, and let us know that his screenplay for *The Molly Maguires* was a metaphor for informers like Elia Kazan and others who had squealed to the House Un-American Activities Committee.

Sean Connery was quite the reserved type until he decided you were safe enough to talk to, but turned out to be one of the best storytellers of all, especially about his early days in choruses of *South Pacific* and other musicals. He had a good self-deprecating sense of humor. He was married then to a rather fiery actress named Diane Cilento, and they had a small son named Jason; they arrived to join Sean sometime after the film began.

Although provided with a driver, Connery always chose to drive himself. One morning, he arrived on the set with the windshield of his car smashed to smithereens, and he went on and on about how a body can't even park a car in a rural area these days without some goddamn vandal doing it damage, and if he caught whoever it was they were going to have their asses kicked good and proper.

We all made the usual murmuring sounds of sympathy and went about our business. Came lunchtime and we were joined by little Jason Connery, who, when we were all gathered around the table, piped up in his English accent, "Daddy, Mummy was so very angry to break the window in your car. Why did she do that?"

Sean whisked the lad away so fast we never heard the rest of the story.

It was altogether a pleasant summer, as Diana and Conor were able to come and live with me in a rented house. Some members of the company had set up a daycare center, so our lively three-year-old boy was kept occupied.

When you are on location with a movie company, there is a womblike quality to life itself. The film becomes the whole of your existence, and when you have a reasonable part you are well taken care of. They drive you back and forth, they feed you, they clothe you, all medical needs are met, with the result that you shut out the world and all its turbulences and troubles because you are too occupied with wondering if your closeup shots are going to be in the final cut.

That year was a strange one, with tragedy—personal, national, and international—hitting everywhere. My sister-in-law's husband, Warren, had returned from Vietnam safely and had gone to work at NBC. On the surface he was still the same ebullient and cheerful lad who went on that foreign venture, but a series of car crashes while drinking belied that. A researcher at NYU later came up with the statistic that veterans of the assault on Vietnam had proportionately more car accidents than any other group, and Warren had several. The last one occurred when riding with his brother Jimmy, who survived, but Warren suffered injuries that led to swelling of the brain, irreversible coma, and, after a few weeks, a merciful death.

For a while, though, I was seduced by the comforts of the movie actor's life and abandoned all protest and demonstrations.

When our location shooting wrapped at the Pennsylvania site, the company and all our families were airlifted to Los

Angeles. We checked into the Beverly Wilshire Hotel, but that proved to be disastrously expensive, as we'd no kitchen facilities, and the per diem could not quite cover the expense of room service. Diana, we had discovered, was very much pregnant again, and all this gadding about far away from home was a bit too much for her, and getting settled was essential. As small as my part was, I still had a couple of months of shooting to do.

Despite my being in many scenes, I had only one line to speak in this entire epic, and a memorable line it was. When the informer James McPartlan appears in the village, he is understandably viewed with suspicion. The cover story he presents is that he is on the run from the police, for "pushing the queer" in Buffalo. No sir, that is not an assault on a homosexual in upstate New York; it was the jargon of the time for passing counterfeit money. When this sham criminal queried me on getting lodgings, I responded with, "There's a train leaving in twenty minutes."

There's another scene wherein the police raid my saloon and a magnificent brawl breaks out 'twixt the miners and the constabulary. At one point, the chief of police whacks McPartlan on the head to make it clear they considered him part of the mining riffraff and throw off any suspicion that he was their spy.

Harris and Connery, who was also in this scene, went to Marty Ritt and asked him to postpone shooting the fight; they were doing all their own stunts, they explained, and might get hurt, and it would be better to do it last, just in case. Marty said that was good thinking, and added, "By the way, I like Malachy. Let's keep him, too."

I stayed on the production until the end, and that's how I came to be one of the highest paid one-line actors in movie history. As Connery pointed out, "If I were getting paid as much as you for each line, I'd never have to work again."

So it was that Diana and I were looking at a splendid house in

Los Angeles that had once been owned by Will Rogers. The real estate agent told us it had last been rented to one of the Beach Boys. He'd let in a bunch of squatters and then abandoned the place, and it was now up for rent again. It sounded great, as it was furnished and the price was reasonable, and the only problem would be kicking out the squatters and getting some bed linens and a few other assorted items along those lines.

When we arrived for our inspection, we had a bit of difficulty getting in. Inside the house, a dozen or so young people were lounging about, with a good strong smell of pot permeating the whole place. They were not a friendly group; indeed, they were quite hostile in demeanor, which made me a bit uneasy, and made Diana downright fearful. Then we went out back to look at the swimming pool (in the shape of the state of California), which the real estate agent had warned us might need a little work. A little work, indeed. The bloody thing looked as if it had not been cared for in years, and it was half full of scummy water. However, that was not the focus of our attention.

Seated at the edge of the pool, his feet dangling, was a thin, smallish, bearded young man, dressed in a black shirt and jeans. We said an awkward hello, but there was no response from him other than a savage, hate-filled glare. Diana began making frantic signs to me, indicating that she wanted to get the hell out of there. I told her to relax, there was nothing to be concerned about, that I would take care of everything.

When we left, I told the agent we'd take the place.

Diana took me aside and said, "If you do, you will live in that house all by yourself, because I wouldn't live there if it were rent free." We continued to argue about it, and I eventually told the agent that I'd changed my mind.

Diana had told me that she had never in her life felt such malevolence before, and that she felt the people in that house were a threat to the baby she was carrying. I, being full of bravado, was going to boot them all out and take over the manse, but was glad I hadn't tried. The assorted gang in possession of the house were the followers of Charles Manson, the young man with hate-filled eyes seated by the pool.

We eventually found a reasonable house in Brentwood, complete with garden, and I settled in for the remaining two months of the shoot. I began commuting to New York once a week to tape my new television show, *Sound Off with Malachy McCourt*, but the first show was taped in California, with Sean Connery and Richard Harris as guests. We shared a bottle or two of the whiskey beforehand so the wheels of talk wouldn't squeak, and though I was a bit nervous with first-show jitters, it all went smoothly, more of a conversation than an interview. Harris and Connery didn't let me down at all, as they were most supportive and quite entertaining.

I did my best to entertain, as well. At one point, I asked Sean Connery what I termed a "pedestrian question," to wit, "How did you get started in showbiz?" Sean said he was playing amateur soccer and doing amateur theatricals, when he got an offer to act for money and an offer to play soccer for money. "I had to make a choice between becoming a professional soccer player and becoming a professional actor," he told me.

Seeing my opportunity, I asked, "And what did you decide, Sean?" Connery began to answer, and then realized he had been sandbagged, recovering his equilibrium quickly enough to join in the laughter at his own expense.

My single spoken line in *The Molly Maguires* was scheduled for the second-to-last day of shooting, and a grand day it was. Harris, as McPartlan, entered the soundstage re-creation of the Pennsylvania saloon and asked about getting accommodations, and I delivered my line:

"There is a train leaving in twenty minutes."

"Print it," Marty yelled. "Now, reverse shot, with camera on Harris."

"Print that one, too," sez Marty, after we'd done it again.

He chose to use the second take in the end, and my face was not seen as I spoke my single line, and thus a great closeup was lost to cinematic history.

The film was released to a good gush of publicity, but it wasn't successful. It's possible that hanging the hero, whilst the informer walked away smirking, didn't appeal to the average moviegoer. Nonetheless, the film has acquired a bit of, as they say, cult status, so it can be seen on television regularly.

It had been a marvelous five months, and I was delighted to have been part of such a congenial company and a bit downhearted at the ending of it all. However, I still had the television show and I was reveling in my new role as host. We shot in front of a live audience, and I felt as much like a host with guests in my living room as I did a talk-show host. I liked to walk down into the audience to talk to them during the show, and I think I was the first talk-show host to use a shotgun mike Phil Donahue–style, extending it to someone in the audience so he could speak his piece on the air.

One of my guests was Betty Friedan, the author of *The Feminine Mystique* and one of the early leaders of the women's liberation movement. At the time I wasn't too enlightened on feminist issues; if I look back on it now, it's hard for me to believe I held such Neanderthal views. There I was, wildly liberal on the war, on labor unions, on every social and political issue you could name, but I held the traditional patronizing, conservative attitudes on anything having to do with women's issues. The other guest that night was the actress Pamela Mason, an acerbic Brit married to James Mason, who was anti–women's liberation in all its forms. I retain the distinct impression that Friedan showed me the error of my ways and managed to put Pamela Mason in her place, too.

Muhammad Ali, in a total professional eclipse at the time, due to his refusal to have anything to do with the war in Vietnam, was another one of my guests. Before we got to the controversial stuff, I asked him if he wanted to talk about his Irish heritage.

"What Irish heritage?" sez he.

"Your grandmother O'Grady," I told him, and, indeed, I had done some research, and Ali had an Irish grandmother.

He was quite surprised I knew that, and then I proceeded to tell him what part of Ireland she came from, something he'd never known.

"You know more about me than I know about myself," he said, smiling.

I went on to ask him how he justified the violent nature of boxing when he refused to go to war.

"When I get in the ring," he told me, "I get in the ring with someone who chooses to be there. We're two equally trained fighters. We both have fighting skills. We both asked to be there. I'm not shooting down little brown people on the other side of the world who didn't ask to fight and don't even want me to be there."

I didn't see how you could argue with that, but the floodgates of hate opened after our chat on the air. Just because a principled man refuses to travel ten thousand miles to kill people in a jungle—people whom he didn't even know, I might point out—

and because I allowed him to state his reasons for same, myself and the TV station were subjected to a wave, a veritable tsunami, of outraged calls and letters containing threats and vows of vengeance, all in the name of love of country. However, Ali is still alive, as is McCourt, while thousands of Americans and Vietnamese are dead, as a result of somebody's patriotism.

The viewing public responded with another outburst of outrage when I invited some students from Columbia University who had occupied the dean's office to appear and explain why they thought it more important to continue their education than to travel far away and kill strangers who had never done anything to bother them.

On the other side of the debate, I had as a guest that famous old windbag, George Jessel, the self-appointed Toastmaster General of the U.S.A., a crashing, thunderous bore. Along with the jingoistic moneybags Bob Hope, he traveled to Vietnam to entertain the boys before they were scheduled to be killed. George claimed this tired road show made him eligible for a Purple Heart, which was a lot of old rubbish, as only members of the armed forces are eligible (even though the late unlamented Senator Joseph McCarthy got one for falling down a companionway drunk while on his way to the Pacific war zone).

George must have had his sense of humor removed in a field operation, because he displayed none at all on my show. Lacking the wit to respond to my comments on patriotism and the war, he decided I didn't appreciate the great favor he did me by agreeing to be there and stalked off in the highest of dudgeon, leaving me with about thirty-five minutes to fill. Fortunately, 'twixt the audience and myself 'twas no problem.

On my birthday, twentieth September 1968, I was getting ready to do the commute to Manhattan, as I was still winding up a few items in L.A. I called the producers to inquire about the guest list for the next show, and was informed that there was no guest list, and I was no longer a television host, as WOR-TV had canceled the show. That blow left me in a calm, frustrated rage. I said, "I see," and went off on a toot to drown the sorrow of losing my niche on television. In my own head, I had been a splendid and witty television host, and I had foreseen the day when I'd go national and be the equal of Jack Paar, Johnny Carson, Merv Griffin, Dick Cavett, and Mike Douglas, but that little ambition was nipped in the bud.

I had also looked forward to the financial security of the job. Diana was pregnant with Cormac, Conor was a happy three-year-old, Nina was doing well in the group home in New Jersey, and whenever I saw them, my other children, Siobhan and Malachy, seemed happy enough with their mother. But just when you have life ordered and all the participants in it in their rightful place, down comes the divine banana peel and you end up on your arse.

A few years after that episode I was in the Shelbourne Hotel in Dublin. Standing at the bar not far away from me was an American chap, a little drunk and a trifle too loud in his comments to all and sundry. He spotted me and called me by name, which surprised me, as I didn't know him. Then he asked me if I knew how it was I got fired from my television show, and proceeded to tell me.

"You see, McCourt," he said, gleefully, "there was a bunch of us at the New York Athletic Club didn't like what you stood for. Didn't like you, or your show, or your left-wing bull-shit, and we didn't care for the disgrace you were bringing on the Irish. We knew the O'Neills, the owners of your station, so we called them for a meeting and told them to get you off WOR or else there would be repercussions in the business community, especially amongst the Irish." He beamed at me, smug.

"So that's what happened," sez I.

"Yes! That was it," sez my friend. "But if it's any consola-tion, the O'Neills didn't know you were dishing out your com-mie bullshit because they're in Boston and hadn't even seen the show. Soon as we told 'em, though, it was curtains for you!"

"Interesting," sez I.

He clapped me on the shoulder. "Can I buy you a drink?" sez my righteous friend. Seemed as if he thought I should be just as delighted as he was at that bit of information.

"Indeed you can, and a couple of them, as I need them after hearing that. I'll have two large whiskeys together in one glass, no ice and no water."

The busy barman served up a drink to my new right-wing pal and the double large whiskey to me. I toasted to America and freedom of speech and, with a swift motion, threw the drink in his face and smiled. Small satisfaction against the ris-ing tide of Irish conservatism and respectability, but it was something.

The gloom of a five-o'clock shadow had enveloped the land in the person of Richard Milhous Nixon, who was going to tough it out with thecommiesthepinkostheperverts, the un-Americans who were the backbone of this traitorous antiwar movement.

'Twas a bad time for liberals, as all the ills of the country were blamed on those of us who felt that we are entitled to life, liberty, and the pursuit of happiness. The Mad Manson had his "family" kill Sharon Tate and friends one night, and the following night slaughter Mr. and Mrs. LaBianca, and all over the country the conservatives said, "This is what it all leads to!"

I'd met Sharon Tate and her husband, Roman Polanski, at a dinner given by Mia Farrow, and they were as excited as any couple about the arrival of their first child. I, being an expert on everything, gave them lots of unsolicited advice. But they never got the chance to take it or ignore it. Horrible details of the slaughter leaked out; Sharon Tate and the baby inside her stabbed so many times they could hardly count the wounds. I still get the shudders when I think of what might have happened had I forced them out of the Will Rogers house.

Violence was erupting everywhere in 1968, and I saw it all on television: the killing of Martin Luther King, Jr., and of

Robert Kennedy a few months later; the beatings of students at the Sorbonne in Paris; the Soviet invasion of Czechoslovakia; the police gassing and clubbing and dragging through the streets of the antiwar demonstrators at the Democratic National Convention in Chicago.

It all fomented a raging helplessness in me, and like a typical alcoholic, my reaction was to drink more and act more devil-may-care, as any self-respecting rebel should.

Cormac St. John McCourt was born on December 16, 1968, a vigorous, healthy, strong-voiced little lad, and there I was, out of work and fearful of having to go back to the bar business after my small fling at stardom. After all, what would They, the ubiquitous They, say at finding me at work in yet another saloon. Back in New York, I managed to get some commercials, and the residuals from these ventures kept me afloat, but only barely.

Of course, I was out gadding about, pretending that the money was no problem, but the reality was that the rent went unpaid, we were frequently without the use of the phone, and were often near enough to having the utilities cut. Still, Malachy went about New York proud as can be at his capacity for whiskey, for talk, and for picking up the bill, or at least charging it and tipping outrageously. Sean O'Casey said, "Be brave, be brave, but not too brave lest it turn into bravado." I must not have been paying attention that day.

In those days my own mind was my own worst enemy, as it presented the world to me as a mocking, dangerous, and demanding place. Though I could move smoothly and easily through most human situations, dark little demons were always gnawing at my soul. There were too many times my mind went back to the horrors of the hovel I grew up in, in Limerick. In the recesses of the soul I was of the opinion that I wasn't of any real worth.

Growing up in want had left me marked in ways both overt and covert, as I suppose it does anybody who grows up in such circumstances. I was always glad to get work of any kind that kept me from exposure to the elements, and I waved away discussion of money. Value, worth, and price were just a jumble of words in my mind designed to embarrass me.

Like many others born in poverty, I'd find myself prowling our apartment and switching off lights, pulling plugs on electric clocks, growling about profligacy and waste of money and resources. When it came to food, I wasn't as concerned with quality as with quantity, and consequently there was a lot of spaghetti and meatballs, potatoes, and rice and fillers served at my insistence. I didn't realize there are other hungers besides physical hungers, that these were spiritual and emotional hungers, but it would be years before I knew that. Those were what was really driving me, but the memory of want and longing was always there.

The resources, financial and otherwise, had descended to a very low ebb, causing the bill payments to be very tardy. There were irate calls from landlord's reps, and smooth but threatening calls from the phone company, and of course the light and gas folk, and anybody else who'd been in my economic orbit. It seemed to me that if I left two bills together on the desk, they had intercourse and produced a litter of other bills, with the result that the house was aflutter with a snowstorm of threatening collection notices, and the phone calls. I almost went back to my old tactic of stamping all the envelopes containing bills DECEASED: RETURN TO SENDER, but of course that wouldn't work for very long.

Representatives of capitalistic enterprises such as utilities and real estate assume that all of us share their awestricken view of the sinfulness of not forking out the readies when due or demanded. I got quite good at the diversionary chat with these callers. Philosophically we had different ways of looking at the subject under discussion. I had been taught that the most feared possibility in my life was the prospect of spending eternity in the snarling company of Satan and the millions of vicious little devils armed with pitchforks. But for all that suffering, I found the factotums of my creditors wielded a threat designed to plunge me deeper into fear and compliance than the Church had ever imagined, to wit, damage to my credit rating.

You may whip your child or beat up the elderly, but nothing, they thought, would leave you facing the world in such miserable, bedraggled, head-hanging shame as losing your credit rating. Bring me a loaded gun and show me to the Golden Gate Bridge: I was reduced to paying cash for the essentials of life, rendered an object of scorn and derision for holders of gold cards, platinum cards, and diamond cards.

The landlord's representative disliked tenants with children, tenants without children, the maintenance staff, and anyone who entered the building. Her idea was that all of us should move out so that this building could be kept in pristine condition and, sharing her passion for bricks and mortar, we would be allowed the privilege of paying rent in absentia. Present or not, my rent was generally in absentia for several months at a time. She would get on the blower and in a tone hurt but harsh demand to know why the rent had not been paid. My response was always the same: Because I don't have it. She would inform me that they were trying to close the books and I would respond that all she had to do was grasp the right-hand side of the book with the right hand and the left-hand side of the book with the left hand and bring them together. The humor and practicality of that approach escaped her Teutonic mind.

Nor did she appreciate the wisdom in my declaration that the fact that I owed her outfit some money was their problem and, as I had a sufficiency of problems in my daily life, I'd appreciate it if they wouldn't use my telephone to tell me of their problems.

Nina, my stepdaughter, was doing well in her group home, but we were stunned by the thunderclap announcement that it was closing due to the illness of the owner. Once again began the quest for a place that would accept her. Facilities for the retarded in the U.S.A. "want the brightest and best-functioning folk," and many places wouldn't take Nina because of her low-functioning abilities and the low so-called IQ.

After a long, painful search, and some financial help from Diana's uncle, we found a place in the Westchester area run by an ex-Marine. Nina was accepted on a probationary basis, and once more we packed up her belongings and once more with trepidation we delivered her to her new home, a nice large house with about twenty-five other folk.

With the gloom of debt enveloping my life 'twas hard to be a loving husband to Diana and father to Conor and Cormac. I was preoccupied and didn't pay them nearly enough attention. My concentration would wander away from them when I was with them, and I was seldom playful; after a while, I'd be barking at them again. I was much too focused on what I didn't have to appreciate what I did.

Because of the dearth of acting jobs, I had to cast about for the source of the weekly income and the kindly John Cobb gave me sustenance by putting me behind the bar of his new hamburger joint in the financial district. It was quite the eclectic clientele: young brokers of all stripes, fellows from the floor of the stock exchange, some bankers, and a plethora of construction workers, concrete pourers, lathers, ironworkers, and laborers all involved in building the World Trade Center. (To satisfy Nelson Rockefeller's Edifice complex, as was said at the time.) All of these various sorts were of a conservative bent, a bit suspicious of me because of my longish hair and flowing beard. I was able to mitigate their fears by the fact of being Irish and keeping my political opinions to myself most of the time.

In those days, people said, "Don't trust anybody over thirty," but even though I was approaching forty, I went out and marched. We marched against the war, and rallied for Eugene McCarthy. We marched and sang and drank and drugged in a frenzy of new liberations. Everywhere we demonstrated there was the danger of being water-hosed, bitten by police dogs, gassed, or shot. I was pushed and shoved on occasion and did get a whiff of the tear gas once or twice, but that is about as far as it went with me.

David Hess, who wrote a number of songs, including "All Shook Up," was a young fellow I'd met playing rugby and a regular at my bar, Malachy's. At this time, he was doing the producer bit at Mercury Records, and he came up with the concept of myself doing an album of songs, poems, proverbs, and observations. Why he thought this was a good idea, I can't say to this day, but who was I to say no? He secured the services of one Lor Crane to arrange the music, and all the technical stuff was arranged by Jack McMahon. Though I'd been on radio and television, this was my first encounter with a recording studio. The music was prerecorded and I was fitted with earphones and told to sing. The songs were in large part written by Hess and, as my ear for tunes needs a lot of work, the terrified croaks that emerged from my throat put all the pigeons within a mile radius into startled flight.

Though the most noticeable qualities I brought to the project were a lack of musicality and a fear of failure, David guided me through the process, and an album was made. It was entitled *Malachy McCourt and the Children Toll the Passing of the Day*, and was launched and immediately sank, and has not been heard to this very day. However, the album has become something of a minor collector's item, and it has been spotted at the flea markets with a twenty-five-dollar price.

Despite my lack of success as a recording star, I was not without preparation for the calling. As run-down and deprived a place as Leamy's National School in Limerick was, we did have the bit of singing led by Mr. Tom Scanlan, pumping away on the harmonium. Mr. Scanlan was a better sort than most of the teachers there, who were as a rule as beaten down by the poverty of their students, and the lack of prospects for us or them, as we were ourselves.

Scanlan, for some reason, had yet to become mean and bitter. He looked past the rags we were dressed in and took an unmistakable joy in our singing. When our voices came together just right, he would smile, and we would smile, and for a little bit the poverty and despair of growing up in the lanes of Limerick would disappear. We sang songs in Irish and plain chant in Latin, not knowing what in God's name we were saying, but singing heartily nonetheless, and we sang patriotic songs in English, and Mr. Scanlan was proud of us.

As ragged as we were, we got cleaned up, shared some boot polish with each other, and got to sing at a concert with grown-up performers, at the Mechanics Institute, a large hall having something to do with unions, which served as a community center. I was about eight years old then, and it was a grand place to our ragged young selves. Jackie Adams, my friend and first drinking companion, was the star of our choir, as he had a great soprano voice. After our three songs, the place was rocking with applause and we were told we could watch the rest of the concert from the rear of the theatre as our reward. I'd never seen live people on a stage before—singing, dancing, playing musical instruments, reciting—and it seemed magical to me.

We were asked as a result of this appearance to sing at another concert, this time at an even bigger and fancier venue, the Lyric Cinema. Again the arse of my trousers was patched by the mother, and shoe polish was applied. The stockings were pulled down to conceal the holes in the back, the hair was given the cow's lick, and off I went, shiny-eyed and barely breathing with anticipation. The mother couldn't afford the shilling to get in, so there was no admiring family circle to spot me in the second row of the choir, third from right. We sang and sang our four songs, once more to great applause, and we were told again we could see the rest of the concert for free, but had to go out the side door and then go around to the big doors in the front for admittance.

The big doors were shut and immovable, so a couple of the lads knocked. There was no response, so we pounded a bit louder on the wood. The door was opened and a face with two teeth in it, one up and one down, like transplanted tombstones, said, "Shush and go 'way."

"The man said we could come in if we came to this door," we chorused.

"Sha," said the face. "Ye are a pack of lying blaguards, and ye are not welcome in yeer own hovels, so how can ye be welcome at the Lyric?" and he slammed the door. For a moment we were all silenced by the treachery, the perfidy, getting us out to the street by duplicity, and then calling us names. We pounded and kicked the huge doors trying to hold back the tears of rage, humiliation, and disappointment, but there was no budging it, and as much noise as we made, we knew the doors were so solid and far away from the stage that there would be no interruption of the concert.

As the lot of us, thirty boys or so, trudged away from the theatre, we came across a construction site using great big planks as barriers. Our minds were as one when it came to recognizing weapons against authority, and 'twas the work of a mo' to grab a big weighty plank and carry it back to the Lyric. On the count of three, we charged, aiming to smash down the formerly impregnable doors. Our work was going just grand, not a guardian of the law in sight, and we suspected weaknesses would appear in the structure any moment, when the big door suddenly opened wide.

In the midst of another exhilarating attack, we went charging through the door, only to be stopped by a wall, a solid wall, causing a huge disassemblage of battery rammers. Looking up from the floor where we'd tumbled into a heap, we saw Mr. Scanlan glaring down at us, his tasher aquiver, demanding to know the meaning of this hooliganism.

We shouted all at once our tale of the treachery that had befallen us, and Mr. Scanlan promptly lacerated the management for what they had done and then marched us into the theatre, where we sat on carpeted stairs and were once again transported by seeing live people performing on stage.

We never sang in public again, but we continued to sing in school, and I suppose those memories had something to do with why I took to the stage when I got to New York City.

It isn't often that poor, powerless people, especially children, score a victory. People seem to think that because you are a child and poor, you have less sensitivity to the frustrations and disappointments that rise up in your face day after day. You are supposed to swallow your feelings at being dressed in rags and being looked down upon by the shopkeepers and being hungry every day. You were supposed to accept it as your lot, for surely you had done something to deserve it. "Stop your wailing and whinging, you little crybaby," I'd be told, and I would stop crying and whinging, and instead store up in my little rancid heart dreams of revenge and very detailed plans for retaliation. And if I never carried out my plans, I never let go of the humiliations and frustrations.

Even when I felt I was the cock of the walk in New York, I was always on the alert for the slight or the put-down, the sneer at the Irish people, the witless mocking of my accent. At the same time, it was convenient for me to play the part of the wild Irishman, the fellow quick with the tongue and just as quick with the fists, and quick to romance the women.

If ever there was an unexamined life on this earth, it was mine.

A young fellow came to me and said he was interested in getting into personal management and was interested in managing my career. Career indeed! There was no career, but he said he had some ideas. As he seemed an eager and energetic sort of lad, and no one else was taking my door off its hinges, I said fine. In short order he hooked me up with an agent who told me that a local radio station was going "all talk," as he put it, and he would like to submit my name as a talk-show host. I was surprised to find that if you get such a job, not alone do you get to talk nearly all you desire, but they actually pay you money to do it.

An appointment was made and I hied my way down to the WMCA radio station on Madison Avenue to meet a couple of men named Steve Labunski and Ken Fairchild, experts on call-in radio. The format was new to the city, as it had required the development of specific technology to receive, hold, and air the calls from the public and do telephone interviews with the guests.

Labunski was a big, hearty man who looked as if he had never said *nein* to a steak and a stein of suds. Fairchild was a Texan, complete with drawl, but unlike the stereotype, he was a smallish, thin man with a wicked sense of humor and the intelligence to go with it. The first meeting must have gone well, as I had several more interviews with Labunski and Fairchild. I got quite chummy with Fairchild's assistant, Linda Miller, who gave me the wink and the nod that I was in, even though nothing had been said, officially. Even so, there was a nail-biting interval of about ten days before the agent told me I was in: I was to be the Saturday night host from 7:00 P.M. to 11:00 P.M. (Providing there was no Yankee game that night. In fact, my television show on WOR, *Sound Off*, was often bumped off the air by Yankee games as well. If my broadcasting career had its own epitaph, it would read: DAMN YANKEES.)

It was hard to contain the glee coursing through the arteries at the news. Occasionally, though, my exuberance was dampened by the question of what the hell was I going to talk about for four hours at a stretch. Being blessed and cursed with a reverence for words, I approached that radio station on my first day with the heart doing wild, erratic leaps in the chest. Of course, it occurred to me that a few drams of the whiskey would settle the corpus into a more serene state, but another inner voice, one I didn't hear from too often, spoke in the head and told me, "You will bollox this job as you have bolloxed many another job, so don't do it," and I didn't on that first day.

I was assigned a producer named Toby Goldstein, a sweet, bright, sometimes excitable young woman, and in the control room was a very funny young engineer named Jimmy Walker. Tom O'Malley had suggested that Sam Levenson, humorist and author, would be a splendid first guest, and he was. My entry into the world of talk radio was as easy as a slow-motion fall into a cool lake on a blistering hot day.

In the beginning, as in any new enterprise, I felt a great vitality there, an energy, and even though I was a weekend worker, I couldn't stay away from the station. The newsroom was bustling with bodies dashing back and forth 'twixt the telex machines as they clattered out reams of news from Associated Press, United Press International, and Reuters. I got a great charge, a sense of importance, out of talking to the reporters, like Vic Ratner and Steve Powers, both of whom went on to make a splash in broadcasting, and about half a dozen others. Being privy to news before it was broadcast to the general public allowed me to be a trifle more arrogant than usual.

I didn't meet R. Peter Strauss, the boss of Strauss Communications, owner of WMCA, until much later. His father, Nathan Strauss, had founded the radio station in 1922. I'd always thought WMCA had something to do with the Music Corporation of America but discovered it was nothing of the sort. Because WMCA's studios were situated on top of the McAlpin Hotel, where they could stage live music, drama, bands, etc., they just used the first three letters of McAlpin. A trivial bit of information, I know, but I was always interested in the minute details of history, particularly if they affected me personally.

My ears perked up when I learned the somewhat less trivial fact that the Reverend Charles Coughlin, a priest from Royal Oaks, Michigan, had broadcast his programs on WMCA. Coughlin was, aside from his other dark-sided defects, a savage anti-Semite or, to be more precise, a Jew hater. He also hated Franklin Delano Roosevelt, trade unions, and communists, and sharing these opinions in a weekly radio broadcast had made him immensely popular in the thirties. Despite his pro-Nazi stance, in a poll conducted in 1934 Couglin was voted the second most popular man in the United States; the only man more popular was the one Coughlin hated most: FDR. Coughlin was actually slated to run for vice-president on a ticket with Huey Long, but the small matter of his being born in Canada put paid to that little scheme.

CBS Radio had dumped Coughlin in 1931, despite the million letters the network received in support of him, as William

Paley, who was Jewish, was not pleased with the blatant bigotry of this man of God. When the anti-Semite set up his own national network, WMCA was the New York outlet for his flow of toxic sewage. I was fascinated to learn that Nathan Strauss, owner of WMCA, though Jewish, believed in free speech, and would not muzzle or censor even anti-Semites. The jumble of issues in Charles Coughlin's case were all covered by the First Amendment: exercise of religious freedom, because he was a priest; freedom of the press, because he was a broadcaster; the right to petition the government; and, of course, the right of assembly. All of which Coughlin exercised in support of that other lunatic, A. Hitler, who had just taken over the government of Germany.

Coming from the alcoholic patriotic tradition of my father, and having been taught that England was and is and always will be the enemy, and that any enemy of my enemy is my friend, when I was growing up in Limerick, I was in a state of wild confusion regarding our "friend," Nazi Germany. But when the Second World War was drawing to a close, and we began to find out about the inhuman savagery inflicted on Jews, Gypsies, homosexuals, and handicapped people, my revulsion set in. Even though I was twelve years old when it all stopped, memories of those horrible images still loom insistently. I wondered then and wonder now how anyone could have participated in the slaughter of decent men, women, and children.

It was not long before I was assigned the Sunday night 7:00 P.M. to 11:00 P.M. slot on the station as well, and now I had eight hours guaranteed at fifty dollars an hour, so I could quit the bar biz once more, this time, I hoped, once and for all. I was so excited at being an on-the-air host I could hardly wait for the weekends to roll around. Between shows, I practiced smart answers to the callers who were intent on tripping me up with stuff I knew nothing about. There were the people who thought they would get me by saying, "What would you know? You're just a bartender," or "You're not even American. You're just a drunken Irishman." My response there was to borrow a quip from the Duke of Wellington. Though born in Ireland, he regarded himself as British. When faced with the assertion that his Irish birth made him Irish, not British, the Duke replied, "If I were born in a stable, would that make me a horse?"

I made no secret on the air of my anger and dismay at the continuing awful dirty doings of the Nixon administration. Some of my colleagues at the station were cheering them on, with only a couple in complete opposition. Among the Nixon supporters was Bob Grant, a well-known conservative who made his way to New York radio by way of Los Angeles and Chicago. Rush Limbaugh owes a great deal to Grant, who paved the way for his style of radio many years ago.

There is not a single issue that I have ever agreed upon with Bob Grant, and yet I have to grant him the most extraordinary technique I have ever heard. To this day, he begins his broadcasts with a stentorian "Let's be heard! I'm Bob Grant, and this is a program devoted to a free and open exchange of views," and he always ends with, "Your influence counts! Use it!" Of course, his programs were and are far from being an open exchange of anything, because he was hateful to anyone with an accent, exceptions being made for those from Israel. No matter what their subject was, he would, with his great facility for the ad hominem torpedo argument, reduce it all to the fact that the caller was not born here, and thus had no right to speak of injustice or human rights. Then, if the caller tried to return to the original subject, Grant would bellow, "Get off my phone, you creep!" and hang up.

Some people thought they could ambush Grant by being most agreeable at the beginning of the call, and then bursting forth with invective, but he could always tell it was coming, and had them off the air in a jiffy. I learned a lot from that disagreeable man, Bob Grant, mimic, sometime psychic, some-time racist, but the canniest broadcaster in the business, bar none.

Barry Gray was the elder statesman of the bunch, who felt he should talk only to God, or, failing that, only to distin-guished guests in the studio, and he acted highly offended when he had to talk to the general populace, of which, I was led to conclude, I myself was a member. Barry was in the habit of always carrying a gun, ever since he had, long ago, taken the part of Josephine Baker in some racist incident at the Stork Club involving the owner, Sherman Billingsley, and Walter Winchell, and was beaten up by some thugs for his trouble.

Afternoon programs were done by Fred Gale, a very rea-sonable man who believed in vox populi, even if the voxers disagreed with him and were abusive in doing so. Dr. Joyce Brothers did the morning advice-to-the-stricken show, and caused ongoing harrumphing in the managerial suite because she usually gave the full fifty minutes of her show to respond thoughtfully to a single caller instead of dishing out the two-minute bit of advice on how to achieve orgasm with an Italian bread.

Alex Bennett was the up-to-the-minute man-on-the-spot, with all the latest music and conspiracy theories. He was friendly with all the antiwar folk, from Abbie Hoffman to Yoko Ono and John Lennon, and caused more upset stomachs among conservatives than anyone except myself.

Leon Lewis, a genial African-American, did an all-night show, and had people calling in to relate bizarre real-life sto-ries. He played games with them, and managed to steer clear of political or racial discussions. He was a heavy smoker, which gave his voice a raspy hesitancy that many women found attractive. Many people told me they were surprised to find out he was black because, as they put it, "he was so nice."

Jack Specter, a former disc jockey, took care of all the talk about sports: scores, league standings, people being traded, and endless statistics about the Yankees and Notre Dame football, as the station broadcast their games at that time. Subsequently, John Sterling came up from Maryland to be the sportsman. For a brief time, he was always going on about his friendship with that other great sportsman, Spiro Agnew, but as details of the financial dealings that would lead to Spiro resigning the vice presidency began to surface, that talk became muted.

After a time, I found myself sharing a desk with Joan Hamburg, an elegant and energetic woman whom to this very day I consider to be the First Lady of Radio. Some of the other folk who appeared and disappeared were Sally Jessy Raphael and Barry Farber, a proud southern Jewish conservative with a gift for language, a fair man who wasn't afraid to give his ideological opponents some microphone time.

Peter Strauss's wife, Ellen (née Sulzberger, of the *New York Times* Sulzbergers), ran a program called *Call for Action*, which gave government agencies a boot in the behind when a citizen complained of laggardly response to their needs. For a while, there was also a mild, humorless conservative named Jeffrey St. John, who held forth on Sunday mornings after the mandatory religious stuff. His first and only attempt at humor got him fired, when he referred to the Verrazano Narrows Bridge as the "Guinea Gang Plank."

One of the most intriguing characters in all radio had to be Long John Nebel, who'd made a name for himself by talking to people who'd been taken off on round trips by flying saucers and welcoming tales of various other odd phenomena on his show. He married a tall, willowy lady named Candy Jones, and brought her in as his permanent cohost.

One time, during a period when his ratings were sagging, Long John revealed that his wife had been doped and hypnotized by the CIA and sent off on spying missions all over the world, and that she was only then beginning to remember what happened. They cowrote a book about the doings, and got a bit of a flurried interest from the press and a jump in ratings, so, mission accomplished. John Nebel was also famous for the dexterity and enthusiasm with which he delivered live commercials. At times, when he was overcome by his own eloquence, his commercials lasted five or more minutes.

Behind the scenes there were sales folk, who brought in the commercials that inspired John Nebel's eloquence. They took no moral or ideological stance on anything, for their function was to sell air time, in minutes or fractions thereof, to corporations. The airwaves rang with paeans of praise for goods and services, which were often absolute rubbish, and only sometimes anything useful.

They were very agreeable people, this sales force, at the time, all men. If they said, "It's cold today," and you said you liked the cold, those lads would leap right in and say that they did, too, and go on to praise you for being one of the few people in the world discriminating enough to like the cold. They had a stock of jokes—old jokes, hoary jokes, boring jokes— and smiled a lot, and expressed extreme concern for the health and welfare of the person to whom they were selling time. Decent men all, battered by their daily acquiescence, maintained in the service of supporting themselves and a family. A farmer may love the sun and the rain at times, and thoroughly hate them at others, but he is always free to shake the balled fist at a deity and curse as long and loudly as needed, knowing that whether he smiles or scowls, the weather will deliver whatever the weather will deliver. Not so these salesmen, who must smile at their corporate customers, and thank them effusively, whether they were receiving sunshine or rain.

I'd never before realized how transient the world of radio was, with people coming and going, nor how cruel and arbitrary it could be, everything from the rating system to loyal personnel manipulated to suit management needs. There was a gentle, smiling man named Winston "Joe" Bogart, who took care of a lot of the details of running the station, from recording commercials, to scheduling and inserting music breaks, to fielding the phone calls of exploding listeners. Generally speaking, he served as the buffer 'twixt the public and the on-air bods, as well as being a lightning rod for us with management. Joe served the Strauss family for almost thirty years, and was ultimately ill served in return, which treatment was to later have quite an effect on my radio career.

The most successful talk show hosts on radio are toadies to the rich and powerful. They nearly always espouse the conservative right-wing way, tend to be racist, sexist, and homophobic (especially if wrestling with their own inner sexual conflicts), and they're liars and consummate twisters of the truth. But above all, they are patriots. In the theatre of the mind, radio's venue, these patriots paint pictures of a bucolic America, a lost America, where Mom, Dad, Sis, and Junior frolicked white-toothed and happy. Mom cooked, Dad went to the office, Sis and Junior went to school and didn't have premarital sex, and on Sundays they all trooped off, pious but hearty, to the white church on the hill. Over it all flew the Stars and Stripes of the good ol' U.S.A. And, of course, everyone voted Republican.

But then along came the feminists, civil rights activists, homosexuals, and all those dark-skinned foreigners to destroy this idyllic Wonder Bread America.

Because of a firm belief in free speech and freedom of the press—even if you are the only person expressing the dissenting view—I had no idea how pernicious so-called "patriotism" can be in preventing the dissemination of other, "unpatriotic," views, until I entered talk radio. Xenophobic misanthropy and misogyny, and all the other baser aspects of our natures, are routinely encouraged.

I used to receive letters containing nothing but soiled toilet paper; other folks threatened bodily harm, and some nitwit even set the Immigration and Naturalization Service on me. When the INS called and summoned me to their offices to determine the legality of my sojourn in the United States, I wasn't much worried, as I'd been born here, though raised in Limerick. "Come and get me," I told 'em. The man replied that if I didn't come voluntarily, they would arrest me and keep me in jail whilst my case was being adjudicated.

I made the mistake of informing the official I hoped they would arrest me, as it would be a rare case for a man born in Brooklyn, U.S.A., to be designated an alien. And furthermore, sez I, you will make me a very rich man when I sue all concerned for unlawful arrest and detention. Silence. And that ended that caper.

Sometimes, there were threatening phone calls to my house. The children, Conor and Cormac, were scared by the loonies. Diana made a game of it by hanging a shrill whistle on the phone cord, and telling the boys to give a good blast when one of the cuckoos called. That worked well, even if, at times, they would also blast the grandparents and various friends to show off their whistle.

Although I still managed not to take a serious look at myself and my behavior, I began to take the nature of what I was saying and doing on the air quite seriously. I was constantly yelping about the moral and material corruption of Richard Nixon and the war-mongering profiteers, whores, and sycophants who sur-rounded him. There were rumors of government-sanctioned break-ins and burglaries, of wiretapping, interception of mail, harassment, arrests, and false information about antiwar activists and other opposition being authored and leaked to the various media by the CIA, FBI, and other so-called "intelligence" agencies.

I discovered that the loudest and most paranoid of people tend to be those who are most powerless to harm the govern-ment, and therefore are of no interest to the government, the least likely subjects of such nefarious doings. But having loudly demonstrated myself to be sympathetic to their imagin-ings, many of those folk sent me cryptic letters, or called with wink-wink messages about what THEY were up to now, and I wearied of the barrage of conspiracy theory being heaped upon me by these odd folk, even when, as it happened, they turned out to be right.

A very witty and erudite man named Art Woodstone popped into the studio one night to have a chat about his book *Nixon's Head*, a hilarious, wonderfully unkind collection, describing in great detail the gracelessness and solecistic wit-lessness of a vicious and vengeful man. Woodstone and self had a rollicking good time, particularly as we got a grand pro-portion of abusive calls from Nixon adherents. As sarcastic and ironic and insulting as we were, the entire show remained rather lighthearted, and there was certainly never the question of an actual threat to the person of the clamp-jawed one. Nevertheless, we drew the attention of the Secret Service.

I'd been pooh-poohing the conspiracy theorists for a long time, believing it impossible that the government would be interested in a weekend radio host on a five-thousand-watt AM station. But the young Bob Rein, my producer at the time, informed me that I was mistaken, and that I was being moni-tored by several government agencies. I was astonished to learn that the Secret Service had come to the station on the Monday following the show with Art Woodstone and asked for the tapes of my program. At the instructions of R. Peter Strauss, in discharge of his duty as the guardian of constitutional rights, the tapes were handed over without warrant or court order.

After some inquiry, I discovered that tapes of my programs had been requested and handed over several times before, without warrant or due process. When I got on the air the following Saturday night, I told the listeners what was "going down," to use the jargon of the day.

Bob Rein and self and Lula Sheperd, the engineer, were howling with laughter at the total paranoia that resulted. People called in, indignant and upset, many trying to disguise their voices, but most of the time their anger overwhelmed their fears, and their real voices would come through. I was further amused to discover that there is a listing in the telephone directory for the Secret Service. On the air that night, I rang the number, and a male voice said, "Secret Service."

"Are you sure?" sez I.

"Sure what?" sez Secret Service voice.

"That you're still secret," sez I.

He displayed no obvious signs of humor, and asked the nature of my business. "I'd like my tapes back. My name is Malachy McCourt and I work at WMCA radio, and your people have been taking my tapes."

"You will have to call back during regular business hours," sez he.

"You are not listening during regular business hours, and, furthermore, Nixon is not giving up *his* tapes without a court order, so I want mine back."

The tapes were returned on Monday, and as far as I'm aware, they never asked for them again.

Despite the concern of the conspiracy theorists, most people then were like sleepwalking lemmings, rushing to God-knows-where-the-precipice-was, but the country woke up just in time. Ye olde craquelure was beginning to appear in the blue-jawed president's facade of solid power. His operatives got more arrogant, and arrogance leads to carelessness, and then, boom—the sound of locks being picked resounded like the guns of Fort Sumter, and the word "Watergate" entered the daily language.

In public, Nixon condemned any attempts at a cover-up, while in private, he said, "Stonewall it. Cover up. Let them plead the Fifth Amendment, or anything else." Archibald Cox, the special prosecutor, was hot on the trail, and Nixon knew it, so on October 20, 1973, he told Attorney General Elliot Richardson to fire him. Richardson, decent man that he was, refused and resigned. William Ruckelshaus, his deputy, also refused and resigned, so Nixon appointed Robert Bork as acting attorney general, and had Bork fire Cox.

I was broadcasting the night Cox was fired, as this stuff was coming clattering over the teletype machines, spilling out onto the floor, until we were ankle-deep in the effluvium of the "Saturday Night Massacre." As was my wont then, I adopted the detached, nonchalant pose of the professional, though my heart was treating my ribs like the keys of a xylophone as I began my comments on what was happening.

Sometimes the mouth operates without the guidance of the brain, and that eve I began by talking about getting fired, and the various terms used to describe that action, i.e., bounced, getting the boot, discharged, ousted, kicked out, cast off, and then I came to "get the sack," a term used in Ireland and England. "Where I come from, getting fired is called 'getting the sack,' and now Richard Nixon has fired Archibald Cox. That makes him a 'Cox-sacker.'"

'Twas not long, only moments, 'til all the lights of our telephone system began a furious blinking barrage. It should be noted that not a few of the outraged folk calling in used far worse language than I ever had.

On Monday, I was summoned to the presence of Ken Fairchild, the program manager, who, try as he might, could not keep a straight face as he attempted to give me a stern lecture on obscenity, actual and implied, on radio. He finally admitted he was just telling me off because he had been told to do so, as he had quite enjoyed my creative description of our president himself. He suggested that I not stop doing what I was doing, but merely avoid the worst excesses. The mandatory Monday reprimands became a regular thing, but Ken was an amiable sort, and neither of us minded very much.

The meetings might have become a bit more heated had I not been blessed with a succession of fine young producers, who made sensible suggestions when my behavior and language verged on what is known as "intemperate." Though mostly a dweller in cities, I have, on a few occasions, climbed on the back of some docile horse somewhere, and had him amble off and back again. Once, though, I found myself on the back of a galloper. The bloody beast had been frightened by something, and I lost control, and found myself riding at a mad gallop. Much of my time on radio was akin to that—the bucking, the charging, the rearing up and neighing, and, only occasionally, placidly grazing.

So, in addition to booking guests, screening calls, and making coffee, the producers on my show generally kept things running as smoothly as possible by occasionally reining me in.

After Toby Goldstein moved up, Maurice Tunick took over. An unflappable student from the New York Institute of Technology, Maurice had a constant smile on his innocent face, but he was terrifically fast on the expletive button when myself or a caller let go with a bit of blasphemy or profanity. Sometimes he was a bit too reactive, though. Maurice was very conscious of odd people and furtive movements late at night outside on Madison Avenue. One night, as we exited the station after the show, Maurice noticed a parked and darkened car waiting behind his.

"Get in the car fast!" he shouted to me, and I had barely enough time to shut the door, ere my intrepid guardian gunned up the avenue, followed by the ominous car matching our speed. We screeched and screamed through lights, red, amber, and green, and still the pursuers were almost in our trunk. Maurice said, "Let's get back to the station and make a run for it! We'll call the cops when we get inside!"

Back at the station, we stopped abruptly and leapt from the car. Maurice tried to unlock the glass doors to the lobby, but the damn key wouldn't turn. The doors of the pursuit vehicle opened, causing me, having completely accepted Maurice's take on the situation, to nearly chomp down on my heart, which had just jumped past my tonsils.

Of course, when two smiling young girls stepped out of the car, and said, "That was fun! Do you want to go get a drink now?" I felt as silly as one possibly can while still getting over having been scared nearly to death.

On less eventful nights, we'd pop off to some club to hear our engineer, Jimmy Walker, a very funny comic, do his stuff. The late Tom O'Malley, who launched my career when he booked me on the Jack Paar show, was once again booking talent for Jack on a new, but unfortunately short-lived, show. I recommended he get Jimmy Walker on the show, which he did, and when Norman Lear saw him that night, he hired Jimmy for the television series *Good Times*, where he made quite a name for himself as J.J.

One night, as a bit of a lark, I put Jimmy on the air as a

member of some far-out, vaguely violent radical group. He bellowed and ranted and gave vent to a stream of unintelligible verbiage, which made it seem as if I might be in danger of being
assassinated on the air. Some listeners became so concerned that
they called the police. We were all shaken to see a half dozen
police barge in to investigate what was going on, but I shouldn't
have been surprised. The call-in cavalry are the most diligent of
observers. They do not limit themselves to the more glaring circumstances. Whether liberal or conservative, any deviation from
your original, prescribed routine is noted and commented on,
even down to whether you say "Good night" or "Good evening."
They call and complain if you cough, sneeze, sniffle, clear the
throat, or say something ungrammatical, and may Allah assist
you if you yawn. And, of course, they loved to call and comment
on my "accent," which was cause for great merriment amongst
the more ignorant of the species.

But I also gave them much genuine reason to call in. I never,
for example, lost an opportunity to lash out at the Catholic
Church, its opulence, and its hypocrisy in raising huge edifices
to the sky while grinding down the spirit and nature of its
children. I scorned the sentimentality of the Irish-Americans,
who waxed nostalgic about an Ireland that never was and an
America that could never be. People who called in to reprove
me on my lack of gratitude to the beautiful land of my birth
got the full weight of sarcasm and, in retrospect, the rage and
depression I was carrying around without knowing it. One
particularly nasty fellow informed me that he was an Irish
Conservative Nationalist, to which I replied that he seemed to
have a threefold sickness.

Failing to ingratiate myself with that most righteous group,
the Irish community, once again had consequences. I had been
supplementing the income with the occasional commercial
and small parts in films. One of the radio commercials I did,

which looked to be an ongoing source of the readies, was for the Irish Tourist Board, at the request of another bit of a lunatic named Sean Carberry, the man in charge of that sort of thing. After about three airings, when the Hibernian citizenry realized 'twas the voice of the Mad McCourt who was urging all and sundry to go and visit the Emerald Isle, they rose up and unleashed a tempest of fury at the Irish Tourist Board, and, in turn, at Carberry for hiring me, so that little source of income came to an end.

If I indulged unpopular opinions on air, Alex Bennett, the station's resident radical, did so even more. Alex's show was a platform for the antiwar movement and the music of the time, which was interwoven with the politics. What was noise to me—I like melodies that are singable by the average punter— was revolutionary to him. Bennett knew the subtleties, the weavings, the intricacies of the minds and movements of the boyos who composed this stuff, and welcomed them. They, being revolutionaries who imbibed, inhaled, and ingested various relaxing and inspiring substances into their bods, were apt to use expressions on the air that would cause expulsion from any church gathering, and necessitated an almost continuous use of the profanity button.

There is nothing like the energy and psychic and sexual restlessness of the younger folk to upset the conservatives and leave them muttering blanket denunciations of the younger generation, beginning with "In my day . . ." These old farts could talk with impunity about "nuking" people—Koreans, Vietnamese, Russians—by the hundreds of thousands, but should the word "fuck" escape out onto the airwaves, the bosses would hear endless screams of outrage from our nice, conservative listeners. Nice, right-wing, mass murderers won't stand for that kind of language. Fuck, no!

Came the day management decided that Alex Bennett had

gone too far with his antiwar talk and his antiwar guests, and out he went, long-haired and unshaven, on his arse. Out in the street, he joined an ongoing group of picketers, the unwashed, unemployed children of the well-to-do who had chosen this spot to protest all things sacred to Mammon and his minions. Alex and his ouster added one more element to the protest, and soon there were hundreds of the "Hell, no!" generation whooping, stamping, and cheering outside radio station WMCA.

There is a story that an Irishman came upon a fight involving a dozen or more bods and politely asked, "Is this a private fight, or can anyone join?" When I came into work that day, and encountered the demonstration, I found myself picketing the place I worked. At the urging of Alex and my fellow demonstrators, after an afternoon spent among them, I went inside to do my show. This put management and R. Peter Strauss in a very awkward posish, because it would look like they were throttling free speech if they fired me, which they sorely wanted to do.

They just let it ride for the mo', though, and I kept talking, but management has a long memory.

My other left-wing colleague, Fred Gale, who was also on the picket line, got the boot a short time later. A reasonable man was Fred, a man of intellect and compassion, which ill suited him to the inflammatory biz of commercial talk radio. He was fired quietly, without fan- or any other kind of fare. We organized a protest, but it soon petered out, and Fred was consigned to the fate that eventually finds all left-wingers on radio, which is not being on radio anymore.

John Sterling stepped in with a sports show, as they hoped his sports expertise would get great ratings amongst that most sacred group of consumers, young men from fifteen to thirty-four years of age. There is an odd assumption in broadcasting that these are the only folk with hefty disposable incomes, forgetting the millions of widows left well fixed, the older couples, the gay population both male and female. And so the airwaves were filled with young, adolescent, skidding voices calling up to discuss statistics, such as how many times a pitcher spat tobacco whilst scratching his testicles during the fiftieth inning of the third game of the World Series in 1962. It wasn't exhilarating radio to many people, but management put a brave face on it and pretended that this was the stuff and soul of the genre.

Sean O'Casey, in one of his plays, has a character aver that "a bird never flew on one wing." That is not quite accurate, as talk radio is that rare avis that flies on one wing—the right wing.

WMCA's trajectory was a splendid metaphor for the way an ideal democratic society, with free and open expression of opinion, can be plummeted into the Stygian slime, overwhelmed by profit and greed. After a bright, thundering, pyrotechnic beginning, the lights went down, and people of good heart, morality, and commitment were replaced by dull whiners.

After Ken Fairchild left, a series of program directors took over the arduous task of fixing that which had once not been broken. They came in yammering that we were all on one team, part of one family, and that we were going to win the ratings game. Ha! Whenever corporate America yelps about the employees being part of the family, you know that somebody is about to be fired.

I arrived on Saturday evening to do a three-hour stint (as part of the ratings war, they had cut me by an hour), and my then producer, Stephanie Sachs, informed me that the decent man of management, Joe Bogart, had been fired after thirty years of faithful service to the Strauss family and WMCA.

That day, R. Peter Strauss ran into Joe in the hallway and

said cheerily, "I hear you got some bad news today," as if he had had nothing to do with it. Joe had to clear out his desk and office in short order, as he was ordered off the premises forthwith. To say I was outraged would be understating it. Though Joe was officially on the side of the enemy, i.e., management, he was a good skate with a sense of humor who was on our side in his heart.

When I got on the air, I could not or, more correctly, would not, contain the torrents of raging verbiage about decency and compassion and loyalty. Stephanie was trying to shush me in my earphones, but I was roiling, rolling, and roaring on the waves of my own verbosity, and a sense of righteousness drowned any vestige of caution that lurked in the brain. I yapped on about how I was terrified of opening closet doors on these premises lest I find a phalanx of fascists, about the savagery of capitalism and the haughtiness of the wealthy, whose forebears (the Strauss family had once owned R. H. Macy's) had been engaged in the great intellectual activity of sewing bloomers for fat-arsed ladies on Fifth Avenue, and whatever else came into the noggin. I was on the verbose path to self-destruction, and there was no hope or thought of getting off it, as I couldn't wait to hear what I had to say next.

On Monday morning came the call to appear at the program director's office. The person designated to hold my arse to the burning coals was named Ruth Myers. As usual, she proceeded to play the tape of the Saturday night show 'til I asked her to desist, as I'd already heard the show. My punishment was three weeks' suspension without pay, which hurt, as I was quite dependent on that weekly salary. I did manage to annoy the management by applying for unemployment, though. They said I hadn't been fired, only suspended, but no pay is no pay, and I got it.

When I was about to return to broadcasting again, I was

summoned to the office and instructed to say, if asked, that I had been on vacation. That was a bit much for me. While I'll readily admit to anyone that I gave voice to many things loony and vulgar and unconventional on the air, I'd never lied. Not because I'm an inherently moral person, but because I'm apt to blush and stammer when I attempt deceit, and do so even more when I'm caught, so I'd made it a habit not to tell any outright lies. I raised an objection to this edict, but I was informed that if I had any desire to continue working at WMCA, I was to say I was on vacation.

On Saturday night at 7:00 P.M., I plunked the arse on the chair in the studio, as usual, adjusted the microphone, bantered a bit with Stephanie Sachs and Lula, the engineer, and watched the second hand hit the appointed number on the clock.

"Stand by. You're on the air," and off I went commenting on the week's events. Then came caller time, and, of course, the very first query was, "Where have you been for the last three weeks?"

I still hadn't figured out how to keep my job without being a liar, but relying on innate survivor's instincts, I replied, "You might say I was on vacation—you might, and if I say any more, I may be on a permanent vacation."

The caller said, "I see," though I'm quite sure she didn't see at all. The program meandered on, and I did another one on Sunday night. On Monday, I was told by telephone to clear out my desk and hit the road, and thus ended my radio career with WMCA. As word in that business soon gets around that you are an opinionated, left-wing troublemaker, I wasn't offered a job anywhere else, either.

There was the occasional desultory on-air objection, with listeners threatening never to tune in to the station again, which protests were greeted with a yawn and a "Thank you for calling."

That station had already begun its spiral into a boring hole, with the same voices heard program after program, day after day, until the people who fired all the interesting, dissenting folks finally left themselves, when the station reached the bottom of radio ratings. The place was sold lock, stock, and barrel to some religious broadcasters and, if you tune in today, you can be saved if you send in thousands of dollars and accept Jesus as your personal savings and loan manager.

I did broadcast for a while on WBAI, the New York City listener-supported Pacifica station, a very rowdy place, where internal ideological warfare was on the daily menu. While WBAI did not lack for leftists of any stripe, the operation did lack the money for tapes, maintenance, paper, or any other kind of supplies, and there certainly wasn't any money for salaries. The exigencies of making a living forced me to stop my philanthropic broadcasting and see about looking after the pressing interests of an impatient landlord.

I shall always remember 1976 as it was, of course the two-hundredth anniversary of the founding of this marvelous country—and the year I got fired for exercising my constitutional right to freedom of speech. I wish someone had told me how expensive free speech can be.

During my sojourn on the air, Diana was occupied bringing up two very lively boys, getting them to school, and taking care of them after, and doing a fierce amount of housework, all without much assistance from me. It was still important to me to be seen as the rollicking, gallivanting Irishman, the character, the eccentric, the devil-may-care boyo, always on for the adventure, the song, the story. But the very solid defensive fortifications I'd built up around the soul were beginning to shift and open, letting out some toxins and allowing some light to shed a bit of healing. Not that I'd reformed in any significant manner, because I still allowed the drink to intrude on my life, but there was beginning to be a change.

I took my family on some rare and beautiful vacations to Ireland, and several to Cape Breton in Nova Scotia, all arranged by the indomitable Diana, of course, lovingly and enthusiastically, with me giving a superficial hand at the last minute.

On one of the Nova Scotia trips, we had booked passage back on the ferry from Halifax to Portland, Maine. When we boarded, I was out of funds, down to eighteen dollars. I, of course, denied any shortage. In those days my credit rating did not exist, and ATMs were far in the future. We had a bit of food left from our week in a cabin, and I told the children to eat only hot dogs. We had hundreds of miles to travel, and it was fairly clear that we didn't have enough for petrol or food on the way home.

Outside the three-mile limit, there was gambling on this ferry, with slot machines strategically placed about the ship. I popped down to the gaming room and thought I would give a dollar or two to the gaping maw of one of these machines, in the hope of winning enough to buy food for the children. The eternal cry of the gambler. "I only did it to win enough so I wouldn't have to do it ever again. . . ."

Even above the sound of the sea, I could hear the donkeys braying with hysterical laughter as I chucked away a third of the resources needed to get us back to New York City. Six dollars I lost, leaving twelve for the journey. The mindless urge toward instant gratification had won out again, and why I didn't slip into the bar and spend the pitiful remainder on drink remains a small mystery.

A few hours later, on my way to the lavatory, passing the machine that ate my money, I launched into a tirade of curses against this inanimate object. Whilst doing so, though, I rooted in my pockets for a stray quarter. Like a man revisiting a prostitute who has left him infected, I paid for entry again with the quarter I found, and, to my bewilderment, a tide of quarters came tumbling down the chute, two hundred in all, fifty lovely dollars. The grand buffet in the dining room was five dollars for the adult and three for the child, and a feast was had for sixteen.

Gambling could easily have moved in and taken over some territory in my soul had I allowed it, but it seems that dealing with drink, and sometimes drugs, was sufficient to occupy me most of the time. John, a fellow who had worked for me in one of the bars, was as addicted to gambling as anyone ever was to heroin, and would wager on a cockroach race—anything so long as betting was involved. His marriage had been destroyed, his relationship with his little daughter was nonexistent, and he was pursued every day by hard-faced men with baseball bats in their cars.

John told me he would never give up his daughter for adoption by his ex-wife's new husband, who had offered him a good bit of money if he'd consent. No matter if it were a million, said he. It was considerably less than a million when he agreed to sign the papers—two thousand, to keep the leg-breakers at bay.

Another disease stronger than man. I have bought lottery tickets, raffle tickets, handed over dollars for a get-rich scheme, and at other times acquired some superficial knowledge of a stock or two and made and lost money, but that's about it for me. My mother, Angela, on the other hand, was a first-class bingoist. In Limerick, she'd been fond of playing an Irish card game called 45 with her cronies, but here, she took right to the bingo hall. She could hardly remember my phone number, but the woman could play ten or twelve bingo cards at one sitting, the thought of which feat causes me to go goggle-eyed. That was only one of many contradictions she presented to the observer, though.

My brother Mike had gotten himself married to Donna at my house, a very festive occasion, except that the mother Angela did not approve of the new daughter-in-law, because of the fact that Donna had been married before. I winced. Like many a sinner before her, the mother was not averse to heaving the convenient brick at other sinners, which is how she perceived divorced people.

Despite her oft-spoken contempt for Holy Mother Church and her priests, there existed in the woman a despair that not one of us had married a woman who had not previously been married. And just as bad, not a Catholic among them—Protestants and Jews all over the place. Here was a woman who'd been savaged by the Church, had railed against the all-male, above-it-all, ruling clique that ran organized religion, but when it came to her own sons marrying outside of the Church, her nose ascended into the air with as caustic a sniff as could be produced under the circumstances.

When we'd all get together, she'd survey the latest generation of McCourts and say, "Ye can't cross the floor without falling over a little Protestant or a little Jew." Of course, she knew we were doing this just to spite her and make her declining years miserable. She'd sigh deeply the martyr's sigh, light up another cigarette, inhale, cough violently, and act as if God had visited this curse on her alone.

After her exile on Flatbush Avenue in Brooklyn, my brother Alphie secured her a small apartment on West End Avenue, quite near where I was living with Diana and Conor and Cormac. She liked small children as a rule, particularly little boys inclined to be wild, and she didn't mind on occasion taking care of the boys when Diana and I went out together.

Alphie was working at a saloon called the White Horse on Second Avenue, near Eighty-eighth Street, a couple of doors up from Elaine's. (This was a white horse of a different vintage from the other pale steed wherein Dylan Thomas had his last stack of spirits.) Alphie is a well-read, perceptive sort of lad, and can disguise his impatience with the human race. He is also a handsome fellow, and there were scores of beauteous young things who came to gaze at my brother with uplifted faces, watching his every move. If he noticed, he never gave any indication, and, being devoid of the hunter instinct, he was always the sought-after rather than the seeker.

When he married Lynn, his present spouse, the wedding took place at the Community Church at Thirty-fifth and Park Avenue, a liberal sort of place. The ceremony was to be one of those ecumenical jobs with a priest and a rabbi, representing their two nominal faiths, together booting the couple and the assemblage onto righteous paths. However, the Divine One has a habit of putting slippery items in our paths, righteous or not. The priest, Father Jake Jacobs turned up, but God must have told the rabbi to forget it, because that's what he did.

We waited and waited until Father Jake offered to do both rituals. As it turned out, he'd had rabbinical training prior to opting out of the Jewish faith, becoming a Catholic, and joining the priesthood. Before it came to that, though, the crisis was otherwise averted when brother Frank announced that he had secured the services of one Rabbi Bruce Goldman, a bit of a radical who had offended the more conservative elements of his temple and for the moment was practicing his faith in a cab, which he drove to secure a sufficiency of shekels to pay the rent.

There was a bit of muttering when the time to do the vowing arrived and Rabbi Bruce instructed the couple to say, "I'll try," rather than, "I will," but a jolly time was had by all. Or most. The mother's misgivings were not relieved.

Father Jake went on to open a saloon called the Palatine, much to the consternation of Johnny Cardinal O'Connor, Admiral of the Fleeced, and was forbidden to administer the sacraments or do the job of the priest. Eventually, he became involved in the plans of a young computer genius from Connecticut, who talked the good father into fronting a charity, through which the embezzling youth sent soiled and crumpled money, to emerge, pristinely laundered and starched, from the other end. Interesting fellow, Father Jake.

Fortunately, despite the end of my radio career, I did not need to resort to such schemes to pay the rent, as I was becoming something of a regular on the ABC soap opera *Ryan's Hope*.

I began my soap opera career on that show as a U/5 in 1975. There are three classifications of actor on a show like this: extra, under five, and principal. An extra is the body in the background who manages to be seen on camera despite being told to look away from it. The under five, or U/5, is someone who speaks anything from one word to five lines. It could, for example, be the cop who says, "Anything more you can tell me about the perp, ma'am?" Principal actors can speak lines until mad cows are homeward bound. The extra makes about $85 a day, the U/5 about $160, and the principals anywhere from $550 a day to $750,000 a year.

In no time at all, I went from being a U/5 to being a principal, making considerably more money. I played Kevin MacGuinness, an affable Irish barman in Ryan's bar. The character was written as a closet Marxist, a little joke perpetrated on ABC by Claire Labine and Paul Mayer, the producers and head writers. After I'd been on the show for a while, I was offered a role in a movie and left for ten weeks of filming. On the show, Claire and Paul covered my absence by having my character go back to Ireland to take care of his brother's pub, as the brother had broken his leg.

One night, I was in a restaurant in Santa Monica, having the evening repast with some friends. I was approached by a friendly-looking older woman, who greeted me rather familiarly and inquired about my brother. I asked which brother she was talking about, and she replied, "The one with the pub." I told her my brother Mike ran a pub in San Francisco, and that my brother Alphie ran one in New York. She said she meant the brother who had broken his leg in Ireland.

I realized then she was a soap opera fan, and not an old friend of the family's, and explained that the story of my brother's broken leg was to cover my absence while I was away making a movie.

"You are making a movie?" sez she.

"Yes," sez I.

"How interesting! I didn't know you were an actor!" she sez, and off she went. I didn't know whether to be complimented or insulted.

The movie I was out there to do was *The House of God*, based on a novel of the same name, by a doctor who wrote as "Samuel Shem." The movie was supposed to be a comedy, and, indeed, the book was hilarious—treasured by doctors everywhere as the definitive work on the lives of interns and residents at a teaching hospital.

Jamie Cromwell and self played a couple of cops who hung out at the hospital, playing cards for pills. Jamie went on to great fame as Farmer Hoggett in the movie *Babe*. Not everyone in the cast drank and got high, but it was the rare one who didn't. There was never any shortage of cocaine, or any other narcotic, or of whiskey, beer, or wine. Of course, in that state, we all thought we were doing great acting jobs, and convulsing ourselves with laughter at each stupid piece of business. And were we any good? Well, have you ever seen the movie? I rest my case.

It was a first job for one of the cast, Howard Rollins, a fine young actor who went on to star in *A Soldier's Story, Ragtime*, and a few other good productions, but he was dogged by the drugs and could never really get clean. He died of his addictions, tragically early. Others from that film had severe struggles with addictions, but they seem to have overcome them. At the time, though I noticed that others were indulging more than was good for them, I didn't think to consider myself among them.

Not long after I returned to *Ryan's Hope*, Claire and Paul sold the show outright to ABC and relinquished direct control, with the eventual result that it dived headlong from the top of the ratings to down among the most obscure bottom fish. Before that could happen, though, a new nincompoop of a producer came in and gathered the cast together to give us the big speech about how we were all one family and, together, we were going to win the ratings game. As usual, when the corporate types start the horseshit about being the big happy family, you can fairly figure on at least half of those present being dumped in a few months.

Having had a few beakers of whiskey, I, too loudly, said, "Remember, you can't fire your own family." The nincompoop producer smiled with his lips and said he had no intention of firing anyone. That confirmed it.

Before six months had passed, most of the solid members of the cast were gone or relegated to very minor roles. In addition, the producer decided the bar was too neighborly and decided to have the Mafia blow it up. A new, high-tech yuppie bar was created for the show, to serve the snappy yuppies and vacuous young things who made up the new cast and to encourage snappy yuppies and vacuous young things to watch the show.

I was suddenly out of a job without any notice, and I was furious. I wrote a blistering five- or six-page letter blasting the nincompoop who had destroyed what had once been a literate and intelligent show and denouncing the discourtesy of not informing me that my days on the show were numbered. I bunged the letter off to the VP in charge of daytime programming, who promptly passed it on to the ass I was complaining about.

I found out later that he had intended to rehire me after a decent interval, but my letter put paid to that. The moral of the story is: "Burn letters, not bridges."

I did, finally, get back to *Ryan's Hope* some years later, after that producer had left. Then, within three months of my rejoining the cast, the show was canceled by ABC.

I went on to work on other soaps, as well. I was on *Search for Tomorrow* as a guess what? You got it: a genial Irish pub owner.

Just a few years ago, my corpus was rented by ABC to appear on *One Life to Live* as a (do you need a bit of a clue?) genial innkeeper on the island of Inniscraig in Ireland. 'Twas there that the entire population of the mythical town of Llanview, Pennsylvania, the setting of *One Life to Live*, would repair for trysts, for quarrels, to be shot at, to do some shooting. Outside, in the fog and in the mists of the bogs, there lurked shadowy menacing figures, members of an evil terrorist organization called the Men of Twenty One, responsible for every ill that befell the island of Inniscraig. I'm still hoping that when I die God will explain to me the purpose and mission of the Boyos of Twenty and One, because despite my extensive involvement in their doings on the show, I still haven't the foggiest notion of what they were up to, or why they did a single thing they did.

None of that seemed to bother the genial innkeeper, Mr. Kenneally, as I befriended Llanview's Margaret and Patrick, and showed them to their room, all the time blathering on about how those who kiss under a full moon on my island would never part. All my speeches were written in that quaint lingo that non-Irish writers believe to be our way of expressing ourselves. A fair portion of the time I would not have a glimmer of understanding of what I was saying. But press on I would, regardless, because if you have the Irish brogue and you smile a charming smile, there is feck all you can't get away with.

Came the time when these orgiastic overseas activities ground to a halt and all the participants returned to Llanview to continue same back in the Quaking State. Didn't the producer, Susan Horgan, as grand a woman as ever slipped a foot into a Gucci shoe, come to the surface with an astounding scheme of a plot. Why not have the terroristic lads of Twenty

One chase me, the genial Kenneally, out of Ireland, forcing me to set up my hostelry in mythical Llanview. So we did, and I rebuilt my inn, the Wild Swan, in the sylvan glades of Pennsylvania. Fortunately, I happened to have the specifications of the inn handy, down to the scratches on the door where the ould dog would be trying to get out. Of course, this humble innkeeper had the necessary millions to build the whole thing from scratch, as well.

So there I was, plunked down in William Penn's old stomping grounds and didn't it all start again. Outside the licensed premises, the chaps of Twenty One were once again very much in evidence. There were lurkers, lechers, and assassins. There were undercover agents, gunshots, and screeching cars. And, of course, there was an inexhaustible supply of fog and mist, imported, one can only assume, from the Emerald Isle.

Then there was a sudden change in the executive producer's department. The Irish-inclined producer was downsized due to unsatisfactory ratings, and an Irish-disinclined producer came aboard. She decided that all this shite was too hard for the average punter to absorb, and a clean sweep was in order.

Overnight your genial innkeeper, Mr. Kenneally, he of the twinkling eye, warm smile, and charming brogue was transformed into a grim, steely-eyed, black-hearted villain—and the only surviving leader of the Men of Twenty One. Instead of prattling on to Margaret and Patrick about the romantic effects of full moons on lovers, I was now locking Margaret and Patrick in a cellar and turning on the gas. My plan was for the whole Wild Swan Inn to be gently wafted skyward when the furnace sparked. I would then escape on a plane bound for South America, for that is where all villains escape to.

Unfortunately for the newly evil Mr. Kenneally, Margaret and Patrick escaped just before the inn exploded, rushed to the airport, and spotted me. I grabbed a passing nun for a hostage—no villain should be without a hostage nun—who pluckily kicked me in the you-know-what, and I was captured. They led me off in chains and thus ended the tale of Kenneally, his brogue and his inn, and my not insubstantial weekly paycheck. I briefly considered putting an ad in the classifieds: "Unemployed terrorist seeks work. Competent and charming. Best references. All replies to the *Irish Voice*, Box BOOM BOOM, New York."

In 1973, while Diana was busy raising the boys, my daughter Siobhan, then fifteen, and my son Malachy, fourteen, were both in dreadful boarding schools in Florida. Malachy, who had grown into a wild lad, got turfed out. As his mother had moved herself to California, he moved in with us. A very uneasy living it was, too, as he was an angry boy, uprooted from both his school and his familiar neighborhood on the East Side of New York City, and rejected by his mother. Linda had decided she was done with the single-parent thing, and that she was going to live her own life, independent of all familial ties, so off she went to Hollywood, where she changed her name, and has preserved her anonymity ever since.

I was an expert on how to deal with other people's recalcitrant and moody teenagers, but mine own were a mystery. Malachy challenged me on every issue that arose, and it seemed we could not keep from aggravating each other. I kept casting the mind back to my own teenage years to find out if I'd anything in common with this boy who was my son.

I'd not done anything of note in school, having failed the most basic certificate, the "primary," as it was called, and then went to work in a bicycle shop, where I was soon fired for incompetence. A brief stint in an army music school to prepare me for a position as a musician in a military band convinced both myself and the authorities that I had no aptitude for music, the army, or school, and I was turfed out at age fifteen.

I then took off for England and became a houseboy in a Benedictine boarding school, which housed a few dozen Polish aristocrats. There I was, addressing children as Your Highness, or Count, or Baron, whichever they happened to be, which was pretty ludicrous. I was soon fed up with the silly job of sweeping and making beds for the children of no-account counts, but I had to make an adult living wage, as the mother was back in Limerick and money was needed to feed herself and Alphie and Mike. I headed on to Coventry, where I worked as a laborer in the factories and gasworks.

I paid the same weekly rent in the boardinghouse where I lived as the grown men did, did the same jobs, ate the same food. I lived in an adult world, and I didn't have the time or leisure to be a normal, rebellious, moody, adolescent, angry teenager.

The men I lived with in the boardinghouse were all smart people, well read, most of them, even though they were manual laborers. Among them were communists and atheists, and 'twas a real treat to hear them, to find them human and caring for humanity. There were fierce words at times on politics, religion, philosophy, capitalism, socialism. They had opinions I'd never heard anyone give voice to before, things I'd thought about but never said; things I'd never even thought about. I'd listen to one of them, a dedicated Marxist, use his tongue as the surgeon would the scalpel to slice the Church and royal family into slivers to be fed to the dogs. I tried to talk the talk, but I still had enough fear of organized religion in me to think that a pissed-off God might hurl a few thunderbolts in my direction for even listening to these apostates.

Friday being payday, with its little pay packets, the entire boardinghouse crew dropped all differences and adjourned to the local pub. I suppose I could have gone with them, but I was afraid of having to buy a round of drinks, and also, being too young to drink, I didn't want them to see me drinking soft drinks, "minerals," as they were called. So I'd go to the cinema, or to a classical music concert, if the price was right. Once in a while there was an opera in town, so I got to see and hear *La Traviata*, *Il Trovatore*, and the operettas of Gilbert and Sullivan. I didn't know anything about this "classical" world, but I thought I should attend, as doing so would give me an air of knowing something.

Although I didn't raise the boy, I thought my son Malachy might have inherited some appreciation for culture, and take advantage of being in New York City, where every imaginable form of art is available for the viewing, or reading, or attending. But not for my son; he didn't read or go to the theatre or the museum. He just hung out, as they say, in the park with his jailbird friends and their birds. His companions were all thugs, thieves, and cutthroats, with not a redeeming feature to be found in their beetle-browed visages.

I was a glowering inferno in those years, critical, angry, and intolerant of everything young Malachy was. He, in turn, wouldn't come home for dinner or do the assigned jobs around the house and constantly provoked me into explosions of rage. Self-righteous people in the wrong rarely take the time to analyze what they do or say, and thus it was with me. Here I had a wounded and sensitive son who was looking for my approval and love, and all he got was criticism and disapproval of all he was and did, so he turned to circles where he could be admired for toughness and thuggery.

Diana and I began to dread the middle-of-the-night phone calls from police and from hospitals, telling of arrests, fights, and bloody confrontations, but still I ranted, not hearing the voice of the child calling for help. Came the night that the phone rang again about 2:30 A.M., and came the familiar New York policeman's voice, gruff but soft: "Sorry to have to tell you this, Mr. McCourt, but your son has been hurt and is in serious condition at Lenox Hill Hospital. It's best if you can get here fast."

"Why fast?"

"It don't look so good for him. I mean, I don't know for sure, but I think you better get here just in case."

"What happened?"

"He was shot on the street with his friend. I can't tell you any more than that."

I slumped on the side of the bed, trying to slow my heart and keep it from rising into my throat and choking me. I gurgled out some words to Diana, not making any sense at all. I have little memory of getting dressed, and indeed I would not have known if I were dressed at all if Diana hadn't ministered to me and got me to the taxi.

When I arrived at Lenox Hill Hospital, I was told the doctors were still at their business, saving my son's life. The police officers had left, so the information on the shooting was a bit sparse. Siobhan, who was up in New York at the time, joined me at the hospital, and we discussed how to break the news to their mother Linda, in California.

"Just tell her," sez I, "that as it's a bleak outlook, she should get a swift flight."

Linda informed Siobhan that she couldn't afford the fare, and that was that.

We then paid a sneak visit to Malachy's companion, Tommy, who was also shot, but less seriously, as he was shot in the buttocks. A not-too-bright lad was this Tommy. He informed us that he and Malachy were walking along the street and met two other guys outside an after-hours joint. Tommy said Malachy asked them if they had any marijuana to sell, and the response was, "Fuck you!" Malachy said, "Fuck you!" back, and the fists began to fly. Then one of them drew out the gun and shot Tommy in the arse and hit Malachy in the stomach, severing a main artery. If they had not been within a block and a half of the hospital, it would have been death for Malachy, as he had lost something in the region of eleven pints of blood.

The blood bank was alerted, and gave the necessary supplies, but the wound was so deep and so severe that there was considerable doubt if he would survive, and the death watch was on. After hours on the operating table, Malachy was wheeled into the intensive care unit, where they attached all the usual life-sustaining apparatuses to his almost lifeless body, and there he lay, supine, a long, angry stitched slash from collarbone to the pubic bone, bisecting his entire body.

Of course, the tears flowed as I gazed at that savaged skin and, of course, the mind charged back to when he was a little boy who put his arms around my neck, called me "Daddy," and thought I was the strongest and best of all men in this wide world. Of course, I wanted to hug him back to the mischievous, brash little boy he was, before divorce and anger and recriminations soured our relationship. The door of death seems to bring that out in us, and I vowed that should he live, I would make sure to be the perfect father, and make up for all the omissions and transgressions of the past.

A call for more blood had gone out, and friends responded in the dozens—the folks from *Ryan's Hope*, family members, and friends of all sort, including friends from the bars with ninety-proof blood. Soon the blood bank and Lenox Hill Hospital were overflowing with donations in the McCourt name. As the day went on, the calls from friends everywhere kept us busy, going over the sketchy story again and again.

Some of the calls were quite jolly, laden with what passed

for humor, because the *Daily News* had reported erroneously that Malachy had been shot in the buttocks, and his pal in the stomach. After a few of those "shot in the ass" calls, I decided to leave the phone unanswered. As night approached, the good doctor said, "If he lives through this night, there's no reason he shouldn't live to a ripe old age."

I almost prayed on receipt of the news, but having abandoned any pretense of faith in any God, I couldn't pray, not wanting to add hypocrisy to my load of obliquity, and instead said to others, "If you believe in prayer, now is the time to get to work."

During that long night, there were moments when I thought the lad had left for the Elysian Fields, but the expert crew of that ICU made sure he stayed with us. I don't know why surviving a night 'til dawn is more important than surviving any other twelve hours of the twenty-four, except for the symbolism, I suppose, but that dawn was a happy one for the McCourt clan.

When Malachy opened his sleepy brown eyes, we let out the muffled cheer, and that began the process of a very swift recovery. His mother did finally come east, the next week, expecting to find a dying adolescent, but instead found a cheerful if weakened convalescent, who was ready to resume his combative life, so that was not the best of visits for her.

It would be grand to be able to write that all went swimmingly in our relationship after that near-death trauma, but things quickly went downhill again. Malachy resumed his life with the thugs and ne'er-do-wells, and I went back to being judgmental and sneering at all he stood for and believed in.

My mother, Angela, loved that boy, and often tried to soften my attitude by saying he was just like me. My response was that if he were just like me, he wouldn't keep getting caught at his nefarious doings. If he inherited my tendency toward wrongdoing, he didn't inherit the intelligence to avoid the consequences.

So, despite Diana's best efforts to mediate, and some family counseling, Malachy and self ended up nose-to-nose like a pair of raging moose, unable to back down and be reasonable. The final parting came when he was arrested on Fire Island, a resort area near New York, for violating a local ordinance. Once more, another policeman on the phone:

"Mr. McCourt, your son has been arrested," etc. etc.

When Malachy got home, I'd already packed his clothes and belongings. I told him he had to leave, as he was too disruptive to our lives, and out he went.

When I look back on that decision, and my self-righteous "Leave and do not darken my door again" melodrama, I cringe with embarrassment and regret. Malachy was behaving as spirited young men often behave, and I was behaving like a Victorian-cum-Pharisee. He certainly hadn't had the best example; I had done far worse than he ever had, and I was angry at him more for getting caught than anything else. He deserved better. I should have made more of an effort, and if necessary turned to professionals for help.

Thankfully, we're reconciled now, though 'twas many a long year ere it happened. When Malachy left, he went on to make his own life, and in a very New York turn of events, the apartment Malachy was living in converted to a cooperative, and he was able to sell his unit for a goodly sum. Off he went to Australia, New Guinea, Costa Rica, and London, finally settling in Bali, where he runs his own quite successful scuba-diving business, Scuba-duba-do. We are now on the best of terms. Malachy was fortunate; most of his friends from those days are now dead or in prison, including Tommy, who died of drug overdose, leaving behind a young wife and child.

The mother Angela, in her small apartment on West End Avenue, was within striking distance of several bingo halls, where she could indulge her passion. Her radio was always tuned to the call-in station, as she liked the give-and-take of it, and though she didn't read the papers, she was pretty au courant in many areas one would hardly have suspected.

It was evident that she ate very badly, an excess of battered and deep-fried foods, as she was considerably overweight. A gall bladder operation in 1968 had made her stop smoking cigarettes; we had nagged her for years about her short breath and difficulty in walking.

"In the name of Jaysus, will ye leave me alone about the cigarettes!" she'd say. "'Tis the only comfort I have left to me!" She did stop, finally, at the age of sixty-one years, but after smoking for nearly fifty years, she did not escape the resultant horrors of emphysema.

That didn't stop her, though. There was a society called the Limerick Men's Benevolent something or other, a supposedly benign crowd devoted to doing the bit of good. (Warning: Always steer clear of organizations with the word "Benevolent" in their name! If they really were, they wouldn't need be so insistent.) Women were not allowed into the inner circle of the wise bogtrotters and former sheep-shaggers who ran this do-good society. Instead, there was an auxiliary group called the Limerick Ladies, whose function was to provide tea and soda bread for the few functions, and wait on the men. The mother, who could barely light the gas stove, never mind cook, was elected president of the Ladies, and began to receive all sorts of junk mail from various Irish societies, and was invited to various functions and meetings whose agenda consisted largely of planning the next meeting and giving plaques to each other. She did feel honored and important for a while, until it dawned on her that her meager income from Social Security and the bit of baby-sitting couldn't cover all this social gallivanting.

Living so close, we had the mother over for dinner often, and for all the various holidays. I, however, was not the 100 percent loving and considerate son. In many ways I was contemptuous of this woman who was my mother. Because there was a simplicity to her, I found it easy to mark her as stupid, and on other counts marked her malevolent and vengeful.

More than anything else, though, I could not erase from my mind the horrors of a hovel in Limerick, and the mother's foul and putrid liaison with the drunken, slobbering cousin Laman Griffin. Laman, staggering in every Friday night with his own steak dripping blood through its paper wrapping and ordering my mother to fry it for him. We had been taught it was a mortal sin to eat meat on Friday, and yet this grown-up was doing just that in full view of us children: Frank, Mike, Alphie, and myself. We were almost always hungry, and tempted, too, by the delicious aroma as the mother added onions to the frying pan. He ripped the meat with his filthy hands and tore at it with his shards of yellow teeth standing starkly apart in his cavern of mouth.

We had been brought by the mother to live with him, after the father had buggered off, and he berated us for nothing and for everything, for what we had done and for what we had not done. If we forgot for a moment to stay out of his way, it would be a fist to the head or a swift kick, just because we were in reach. Nonetheless, each night, when he came home drunk, we sat up as long as we could to prevent what we knew was coming next. That never worked, and we were ordered to bed anyway. We lay there pretending to each other to be asleep, but in reality wide awake, rigid with apprehension and fear. The muffled voices of Laman and the mother drifted in, followed by the sound of the table being moved against the wall and the chair being placed on it to allow them to climb into the loft above our heads where Laman slept in his springy bed. I never knew if they cared that we heard the sounds of the springs, the groaning, the sucking sounds, the slap of flesh on flesh, or if drunkenness in one and the fierce desire to be desired in the other wiped away any effort to do their fucking quietly.

If she didn't submit to him it's likely we'd be thrown out into the street again, but knowing that never lessened the rage that I felt. One of my fondest dreams to this day remains going back to Ireland and seeking out the man who disgraced my mother. I would slowly and methodically and with deliberated viciousness beat him into insensibility, but I never did that, and most of the time I'm at peace about it.

Those memories were a sickness in the family, and we all carried that rage and resentment within us, too ashamed to talk it out among ourselves. Never 'til the day she died did we, any of us, give her the opportunity, either in anger or in sorrow, to say, "I'm sorry." I found it impossible to forgive her for inflicting the horror of those actions upon innocent children, so my relationship with her remained superficial and cold, always critical and judgmental.

I picked on her, made fun of her, this wounded and bereaved woman, who had started her own life in an abusive family. She had been abandoned by her own father, who buggered off to Australia just before she was born, and someone told me her own brother had sexually assaulted her as a child. With a bitter mother and a tormenting sister as well, 'twas no wonder she married my father. He was a rampaging Irish patriot and a bollox of an alcoholic, but he was not a man to hit a woman, and never raised his hand in anger to us.

She married him, she wrote in a diary we found after her death, because he said he liked her. He was the first person to say that to her. I can't imagine anyone reaching the age of twenty-two years and not having had one person say "I like you."

"I like you," he said. Did he ever say "I love you"?

I never heard her say that, nor him.

When I was a child, I once told her I loved her, and she reacted as if I'd just slapped her in the face, and then she laughed at me. Later, I heard her relating to the neighbors that I'd said this strange thing, and they burst out laughing at this bizarre bit of news. 'Twas many a year before I said "I love you" to anyone again.

In retrospect, I think she wasn't sure any of us should get married because she knew none of us had any preparation for that intimate state and felt that the women would get the worst of the bargain. It came out just the opposite, though, with her disapproving of our spousal choices.

These days, being free of alcohol and other enslaving substances, I can say "I love you" to Diana, which I do every day I'm alive, and there are others in my milieu to whom I can use the word "love" without the usual Irish embarrassment. Diana often talked to me about being kinder and more respectful to the woman who bore me. I wasn't listening or didn't want to hear, because, as I found out later, if you begin forgiving others, you then have to forgive yourself, and to forgive yourself, you need to acknowledge and expose your own wrongdoings.

There was an enigmatic quality to my mother that could never be fathomed by me or my brothers. Angela McCourt was a well-liked woman. People who hired her to look after their children all talked about her magic with babies and her intuitive knowledge of their ailments. She did love the babies, and was patient and loving with them. My children, Siobhan, Malachy, Conor, and Cormac, had the benefit of her ministrations, and Frank's daughter, Maggie, also saw a lot of her.

All of her bingo pals thought she was a paragon of steady wisdom, and good luck to sit with at the bingo table. She had a plump, round face, nice, blue eyes, good skin, no wrinkles, and, generally speaking, a pleasant expression at all times. There always remained a wariness, though, as if she expected to be hit or rebuked for anything she might say or do. At the same time, there could be a certain Zenlike quality to her conversation.

Frank tells the tale of bringing her to his favorite hangout, the Lion's Head, in Greenwich Village, one St. Patrick's Day, when the joint was packed with people trying to celebrate that dreadful day and others trying to escape it. A Welsh philosopher of diminutive stature was putting people to sleep with her theories of the nature of life.

"All things have life," she averred. "Mountains, rocks, the trees, the flowers, plants. They all have life, and you must talk to them."

Turning to my mother, the only person still awake at the table, she said, "Do you talk to your plants, Mrs. McCourt?"

"No, I live alone," replied the mother.

One night, in a dash to her lucky bingo chair, the mother tripped and fell heavily, breaking her hip. She landed in the hospital for the replacement operation, which wasn't success-ful, and thus had to be redone, as the pin had broken. That was when the spirit seemed to ebb from this mettlesome woman. Once she went home, she had to have the attendant come to her house to bathe her and do the laundry, and she was reduced to going about in a wheelchair. Though the dis-ability was temporary, she had been assured, the light of sur-vival began to fade from her blue eyes, and she lapsed into a series of depressions.

The lack of mobility exacerbated her emphysema, the breathing became more labored, until she finally had to be assisted by tanks of oxygen and, over time, hospital stays became more frequent and longer. Her posture became that of despair and hopelessness.

During one of her hospital stays, they sent in a psychiatrist, who diagnosed her mental condition as depression.

"I can't get out of the bed, or breathe, or go to a party," she sez to him. "Wouldn't you be depressed?"

"Would you like to talk about it?" sez the good doctor.

"I just did," sez the mother.

My fiftieth birthday rolled around in 1981, and Diana and some friends conspired to make it a big do. I was delighted, as many of the chums decided to attend. To my astonishment, the mother had managed to get collected from the hospital and brought by taxi and wheelchair to the borrowed penthouse where the party was held. My friend Bernard Carabello remarked to the mother that he hoped she wouldn't die at the party, as he wouldn't be able to stand the excitement. She said she would let him know.

There were speeches and songs and drinks and food, and

then the doorbell rang, and a very pretty young girl announced she had a message and a gift for me. Out of a large bag, she produced a tape player, put it on the table, turned it on, and then proceeded to do the sexiest of dances, and disrobe.

There were gasps and general howls of disbelief. I'm told I turned cranberry red myself. It was quite a modest striptease, really, a gift from Claire and Paul, the producers of *Ryan's Hope*. The dying mother's face was a study in frozen amusement throughout. She told me afterward that she felt she ought to laugh, as the girl was only doing her job.

After the party, she was wheeled back to the hospital and to her bed, from which she did not stir again. The party was her last gasp in the outside world. She was seventy-four years old, and quite sure of her desire to die, and told us so. As she slowly slid toward the grave, we informed the brother Michael that he ought to get on the plane from San Francisco and bid farewell.

It was getting close to Christmastime, and some visitors from her bingo-playing circles tried to cheer Angela up by saying she'd be out of the hospital by the day itself. I had to take this gaggle of elderly women aside and tell them she had decided to die, and they ought not to make her uncomfortable with false optimism. I suggested that they just say good-bye, and let it go at that. Of course, the average U.S. punter has no comfort in dealing with death, so they flee it. I once took a look at the 180 listings for undertakers in the Manhattan classified telephone directory, and not one of them contained a single mention of death or dying. Lots of stuff about final arrangements, and loved ones, but never a direct mention of death. I sometimes think people flock to America because nobody "dies" here; everyone passes on, passes away, leaves us, goes to their eternal rest, breathes their last, shuffles off the mortal coil. But die? Not on your Nellie!

So it was with the mother's friends—if you don't mention death, it doesn't exist.

When the mother demanded they remove the life-sustaining equipment, there was a bit of opposition from various physicians on legal and ethical grounds. After a long conversation with the doctors, I told them, "Don't worry, gentlemen. We come from a long line of dead people."

I doubt that was what did the trick, but eventually all the heroic stuff did finally stop, and the mother seemed at peace. We made our farewells, Frank, Mike, Alphie, myself, and the various grandchildren, and we took turns on the death watch.

At about 2:00 A.M. on the twentieth of December 1981, I was dozing in a chair at the foot of the bed, and awakened to find the mother observing me in a very wide-awake manner, after two semicomatose days.

"What are you doing here?" she asked me.

"I thought you might die tonight."

"I might and I might not, but that's my business, so why don't you go home to your own bed, where you can sleep in comfort?"

"If that's what you want," I answered.

"I do."

"Good night. So, I'll see you tomorrow."

There was no response. I went home to bed, and the hospital called at about 5:00 A.M. to say Angela McCourt had died.

I'd always understood dying people like to be surrounded by loving family members at the solemn moment, but the mother was an exception. I think she would have thought it rude to die in front of someone without saying "Excuse me!"

At that time, Frank was in, as they say, "financially straitened circumstances," as was I, and the brother Michael had spent his spare cash on the flight from California. Alphie was then prospering as the proprietor of a very successful Mexican restaurant, so, when it came to the funeral expenses, Alphie was the man.

The mother's death coincided with Diana's birthday, and two days later was Alphie's wife Lynn's birthday, and needless to say, all celebrations were put on hold. Alphie had planned a sumptuous do at the famous '21' club for his wife, and it was necessary to inform him that the money would be needed to dispose of Angela's body.

Alphie, Lynn, Frank, Diana, and myself all trooped into a local funeral home to discuss the, ah, "arrangements." A tanned man with gold-flecked hair, some gold teeth, a gold tie clip, gold rings, and an enormous gold wristwatch hummed us into a semicircle of seats around his desk. He slid out some well-worn words of sympathy on our bereavement, and then began the laying on of guilt, so that we'd go for the huge notice in the *New York Times*, the enormous bombproof casket lined in expensive fabrics and laces, complete with foam mattress and golden handles, and a hearse and limousines and printed cards.

The man's gold-ringed hands waved in the air as he spoke, as if he'd learned his moves by watching Leonard Bernstein conduct, with the assembled McCourts moving to and fro in unison like a gospel choir. But there's nothing like a shortage of money to bring sense to a meeting, even when there's a surfeit of guilt, and we managed to stop the hypnotic commercial flow and gavel the man to order. As the intent was to cremate the body, I inquired if it were possible to rent a coffin for the doings in the funeral home and return it when we were done. There was a negative response on that one, and we grumbled about buying an elaborate box, only to stick it in a furnace.

At the time, there had been a sanitation strike in New York City, and the streets were quite ripe with rotting garbage. The strike had just ended, and massive pickups were scheduled, so I

queried the man about securing a body bag and leaving the
mother's remains to be picked up on the street.

The golden man had been sitting with his fingertips
together, polished nails glinting, his lips pursed in disapproval
of the turn the conversation had taken. With a heavy sigh, he
excused himself and left the room, leaving a mass of McCourts
unable to contain our laughter, which intensified when Alphie
remarked, "There is a man who's had twenty years' experience in
extreme unctuousness."

When the unctuous one returned, he was determined to
put an end to the irreverence that had been smuggled into his
hallowed halls. After another bit of chat, we were able to settle
on a price for the whole caboodle without insertion of any of
the sly supplementals and irrelevant incidentals.

We decided on the open coffin, because, in the mother's tradition, home in Ireland, she'd be laid out on a bed and the wake
would be centered around her. Much to the consternation of the
funeral home people, we brought in large quantities of beer and
other beverages, and we got a pretty good old Irish hooley going.
At one point, the brothers and offspring all got up in front of the
coffin and formed a chorus line. We sang songs my mother loved,
and we sang songs she hated, and we knew she was truly dead
because she didn't rise up and smite us with her famous right-
hand uppercut. Still, the woman would have enjoyed being the
center of attention.

As people came to pay respects, it was fascinating to watch
them. Having just adjusted their faces to a suitable degree of
lugubriousness, they were suddenly faced with a phalanx of
dancing, singing McCourts, who didn't seem to need sympathy
on the loss of their mother. A bit of solemnity did descend on
us when it came time to close the coffin. I found myself weeping
a bit, and regretting not having taken the time to make some
amends with this aggravating, stoic woman.

"My mother is dead," sez I to myself. "My mother died, my mother will not be seen by me again in this life, and there were things to be said. God damn it! There were issues to be aired, and now she has buggered off to the next life."

Before she'd died, I'd had one small glimmer of remorse from her, when she said there were things she was sorry for in her life. I stopped her saying any more, because I feared she might ask forgiveness for the one thing I thought was unforgivable—bringing us into the home of Laman Griffin and carrying on as she did. I told her to forgive those who had hurt her and, above all, to forgive herself. I got a blank look on that bit of advice, and that was the extent of it.

They put the cover on the coffin, and that was it. A van took her to a crematorium in New Jersey: no hearse, no limousines, no solemn, dark-suited men or black-veiled women dabbing at their eyes with little handkerchiefs. An old bean can filled with her ashes was subsequently sent to the funeral home, where it was picked up by Frank, who subsequently gave them some thought.

There's a certain security in having the parents still living. In a sense, they are in the line of fire, and as long as they are in the front lines, I feel like a support unit. But as soon as they die, I'm moved into a combat unit, next to be mowed down. When the mother died, I wasn't sure whether I was grieving her leaving or my own move toward the grave.

When Diana and myself searched for the residential facilities for Nina, it was with the expectation that we would find a
permanent, loving home for her. A place with the trained people and the facilities to help her acquire ordinary life skills
such as feeding and dressing herself. We found such a place in
Dover, New Jersey, run by a lovely woman named Brunell.
Though it was a severe strain on the available income, we managed to come up with the fees, and all was happy for a while.
But clouds appeared on the horizon, and Mrs. Brunell had to
close down for health reasons, and off we were on the trail of
another happy home for Nina. We found one in Westchester;
not as good, but it was a pleasant and healthy environment,
and again we had the sad task of moving Nina to yet another
strange place.

But every silver cloud has darkness lurking. We were informed
that Nina's school was upgrading, and, as she was not upgrad-
able, they asked us to come and remove her from the premises
within the month. Once more Diana had to initiate the search for
a suitable residence, affordable and understanding of Nina's
needs.

There are times in my life where I am a large, helpless, use-
less hulk drifting in seas of ignorance. I simply don't know
what to do in certain situations, and finding a loving home for

a handicapped child in need of training is one of them. Diana, a loving, determined mother, gets on the path and stays on it until she gets results. In this case, she was stymied, as places were either too expensive, or else they were bare little detention centers where people rocked and rotted their lives and minds away.

We'd been told to look into state institutions. Some said they were horrors, some said they weren't so bad, and being in no position to debate, we began visiting these places. Our first stop was a bucolic-sounding institution, Letchworth Village, with wooded and grassy grounds. Fairly pleasing to the eye on the outside, life on the inside was nothing like life should be. We saw hundreds of human beings, or what had been human beings, crumpled in wheelchairs, on floors, or on hard plastic chairs. One tiny little girl in a wheelchair, who looked to be eight or nine, reached for my hand and wouldn't let go. I asked the doctor how old this girl was, and he said she was fifteen. When the residents don't get therapy, physical and mental, they shrink, he explained. When we got to a quiet spot, away from his colleagues, this doctor quietly and firmly advised us not to even think about putting Nina in this place, not if we wanted her to live.

Our next stop was Willowbrook State School for the Retarded on Staten Island, another institution in a bucolic setting. From their very comfortable and spacious houses atop small hills, the director and his senior staff had splendid views of lovely sweeping greenswards on several hundred acres of wooded lands.

They told us that there was a two-year waiting list, but if we consented to admit Nina through the hepatitis program, she could be placed immediately. As it was explained to us, they were testing a new vaccine, and it was nearly totally effective, except for a few small glitches. They also told us that as 100 percent of the residents got hepatitis, it would be advisable to get this vaccine anyway. What they didn't tell us was that the program was totally experimental, and that the residents at Willowbrook were the guinea pigs. Nor did they mention that the U.S. Army was funding the program.

Yes, parents and relatives did give consent, but as the ramifications of hepatitis and the hepatitis program were not explained fully—indeed, obfuscation was the order of the day—it was not "informed" consent. Of course, nobody would believe that the noble and honorable United States government would ever use innocents in a disease-inducing project. When I asked, much later, why monkeys were not used in the experiments, I was told that monkeys were very expensive.

The setting for the hepatitis program was pleasant enough: Everything was clean, sanitary, and well lighted, and though the first visits were restricted because of the risk of contagion (a minor precaution, they said), all seemed well with Nina. So, we put the blinders on for a while, as we were apt to do when an immediate crisis has been resolved, and went about our daily business.

Except, late at night, I'd awaken to a flood of unfiltered thoughts about Nina in that huge institution, think of how it loomed around her, dominating every aspect of her life. I'd hark back to times when I sat and looked into her beautiful, but seemingly uncomprehending, eyes, and she'd look back at me, smiling, searching my face as a baby does for who knows what.

Does she get lonely? Was there rage lurking in her heart? Did she know she was loved? Can she feel love herself? Did she wonder at all that swirls about her, out of her control?

As the stepparent to a retarded child, I could lay claim to the parental role, but not to the initial grief at the loss of a normal future for my child. Nor was I subject to the guilt and blame that parents, in the despairing agony of giving birth to a less-than-perfect baby, sometimes slash each other with. So I was able to leap, sans emotional confusion, onto the justice chariot at full gallop.

Diana and myself were invited to join the Benevolent Society for Retarded Children, Willowbrook Division, a subgroup of the National Association for the Help of Retarded Children. Both of these groups were moribund and resistant to change. Their main function seemed to be having annual lunches and dinners to honor the self-satisfied directors and commissioners of the various institutions that were quietly and systematically destroying the residents of their hellholes.

But we looked around carefully, and slowly the full savagery and horror of Willowbrook State School began to emerge. We were surreptitiously contacted by some folk who were working at this awful place, and they put us in contact with other parents who had not been brutalized by imposed guilt or the fear of retaliation against their kids. Dr. Mike Wilkins and Elizabeth Lee, a social worker, began talking to the press, though forbidden to do so by the director, Dr. Jack Hammond, a dour sourpuss of a man.

Also leading the charge was Dr. Bill Bronston, a dynamic, intense man, so suffused with passion and compassion that there were days he was so emotionally charged he could hardly speak. Bronston was tenured and could not be dismissed except for cause, but Mike Wilkins and Elizabeth Lee were in a precarious position, as they were not tenured employees and were in danger of losing their jobs.

Ira Fisher, another social worker, took us on a tour of the back wards. When he opened the thick, heavy doors, I was assaulted by smells and sights and sounds that were so awful I didn't want to believe what was in front of me. A look at Diana told me she was stunned by the desperate savagery of this pitiless place, littered with twisted and grotesque bodies, writhing and rocking on floors gleaming with the slime of every excretion a human body can produce. Strange, high-pitched howls and low groans rent the air, interspersed with the dervishlike leapings of jibbering beings. The hard, spare floors and walls reverberated with a deafening, dissonant symphony. Not only were some of the residents retarded, they were driven totally mad by the conditions of their so-called state school.

These "recreational" areas held as many as eighty residents, with perhaps three attendants to administer to their needs. High in the corners of these dank dungeons there flickered the ever-present television, showing soap operas with sleek men agonizing over imaginary lost millions and perfect females weeping over imaginary lost loves. Amidst these insane horrors, with soap operas playing out above their heads, the attendants, no less battered by the conditions than their charges, tried to shuttle and cajole the residents to the lavatory, or to lunch or dinner, which would last all of five minutes.

Our activities were soon noticed by the authorities, and the usual official paranoia set in. Rumors, fueled by the leaders of the employees' union, puppets of the administration, began floating about that a group of communist parents and subversives were going to take over the institution.

Our little revolutionary cadre was plagued by informers, who ingratiated themselves with the institution crowd by reporting on our every move and strategy. We took to meeting at Dr. Bronston's house, inviting only those we were sure could be fully trusted. Among the co-conspirators was one of the bravest people I've ever met, one Bernard Carabello, a twenty-year-old man who had been a resident of Willowbrook for sixteen years. Bernard had been born with cerebral palsy, but misunderstandings and miscommunications caused another tragic miss: a misdiagnosis. Bernard was diagnosed as mentally retarded, and it was forcefully suggested to his mother that she institutionalize him, which she did, when he was four.

As he was considered retarded, the officials spoke openly in front of him, and he fed us information about what was going on inside the facilities. If it had been known he was funneling this intelligence to us, he would have been beaten and put into one of the isolation cells, or they might have designated him a "biter," and, as was done with those so designated, pulled all his teeth. Without the benefit of anesthetic, of course.

Bernard would later go on to become a prominent activist in all areas connected with the handicapped. He earns a good salary and travels extensively hither and yon, giving talks and consulting wherever he is needed.

As Diana and self got deeper and deeper into the morass of slime that was Willowbrook, the rest of our lives were put on hold. Our other children rarely saw us, as we were constantly shuttling to meetings of the Benevolent Society or meetings with our inner circle at the Bronston house on Staten Island.

Staten Island was and remains a hotbed of conservative yahooism, and it wasn't long before the rumors that we were "communists" spread to the outside community. We were treated with hostility and suspicion where e'er we went. Rumors were circulated that there were stockpiles of arms and bombs at the Bronston house, and that in our spare time, all of us commies took part in the postprandial orgies.

As we began to go public, the official reaction to exposure of brutality and savagery was to deny, stonewall, and divert. The word came from on high: "Get rid of the troublemakers." Dr. Wilkins and Elizabeth Lee, advocates for humanity and decency, were fired and ordered off the grounds.

That was the biggest mistake made by the genteel savages who were in charge. The cauldron was boiling and could not be stopped from spilling over. Those of us with media contacts had pleaded with them to please come and look into the hell that was Willowbrook, but they had not yet come. 'Twas hard for them to contain the yawns at the prospect of covering such a story. I had personally contacted the so-called "anchors" of the networks and all of the local stations, but, no, the story wasn't exciting enough. Media types will come to the field of dreams, but don't ask them to cover the plains of nightmare.

Eventually, one young reporter did come, unaware of what he was wading into. Dr. Mike Wilkins had had some contact with him during the turbulent sixties, and now he was a reporter for the local ABC news operation, going by the name of Geraldo Rivera. We met in a diner near Willowbrook and gave Geraldo the keys to the men's and women's wards, but forgot to warn him that after seeing the horrors, hearing the horrible screeching, and enduring the fetid stink, he would be driven outdoors to throw up.

He couldn't commit to getting anything on the air, as he was only a reporter and didn't make such decisions, but when he got the film footage back to ABC, the people in charge saw what was going on under the tender care of then governor Nelson Rockefeller and his minions. They could hardly contain themselves. To say that all hell broke loose would be a classic example of understatement.

By then our group included parents and relatives from all the societal strata, as well as a great ethnic mix. As the get-togethers became more frequent, and we got to know each other better, we all became more open and less embarrassed to talk about what might be. Looking into Diana's eyes, seeing her agony at some of the decisions she'd had to make for Nina, and then the blaze of hope that she could do something to make the world a softer and more caring place for all the vulnerable Ninas, sparked my own hope and love. We progressed from "Let's make the institution a better place" (a dopey idea, we realized) to "Let's have small homes in normal settings with normal furnishings and schools and communities which would take care of any special needs of our folks."

There we were, trying to understand why a God lets handicapped babies into the world and expects us to know what to do about them, trying to deal with our rage and disappointment, trying to understand why other human beings can be so callous and inhuman toward the most vulnerable among us. There were days when I was certain I wouldn't have the willpower to set foot on the grounds of Willowbrook again.

We were obstructed at every level, but we began to prevail. In 1972, a class action on behalf of all the residents was instituted. The Association for the Help of Retarded Children, along with Diana and other parents, were the named plaintiffs, and the ACLU and the Legal Aid Society took the lead in presenting the case in federal court. The director of Willowbrook, a whining, accusatory sort of man who lived in a big, luxurious house like a camp commandant, got turfed out to run another hellhole in Rome, New York. The newly invigorated parents and relatives staged a veritable coup and took over the so-called Benevolent Society for Retarded Children, and so the revolution continued.

Of course, New York State denied there was anything wrong in the administration or in the lives of the citizens interned there. One of the more tragically laughable aspects of this whole period was the bellowing of the director and his cohorts that we were violating the privacy of the folks at the institution, and we should be prosecuted. 'Twas all right to subject these people to every physical indignity they could invent, but, goddammit, you don't fuck around with their privacy!

After two years of marches, and protests, and confrontations with the authorities, it was hard to sit in the courtroom observing proper decorum as the hypocrites and opportunists sat in the witness box and lied. But we had some splendid legal minds on our side, who slowly and methodically reduced the defendants to whining about lack of money and resources. The suit ended with a consent decree agreed to by all parties that Willowbrook residents would be relocated to small residences and, for as long as they lived, the court would look after their interests. No more would we see young people die of pneumonia, the most common cause of death in Willowbrook, because they had to eat fast and got particles of food in the lungs, which caused infections. No more choking deaths, no more pregnancies from rapes, no more amputations because a kid's maggot-infested leg was left in a cast long after a fracture had healed. No more retarded boys and girls goaded into having intercourse for the amusement of bored staff. No more beatings, no more teeth pulling, no more heavy doses of drugs to keep people rocking in semicomas.

It was a historic case, because we got the state to acknowledge that all people had a right to decent human conditions, no matter what their mental status. It would seem obvious to the average punter, but the courts had to force the issue. Similar suits followed around the country.

When the process of closing down that so-called school began, we got Nina into a group home where there were twenty residents in adjacent apartments. It was an improvement, but not good enough, as none of us are at ease living with a score of bodies night and day. So, another move to another group home improved her lot still more, as there were only nine other folks in the home. We can get a subway or bus now for visits, and Nina has mentors who take her for walks to shop and to help with meal preparation and doctor visits.

My involvement in the Willowbrook events was a combination of disparate needs and emotions, a torrent of pity and rage and empathy and righteousness, of drama and love, all surging into a roar for justice, for compassion, and mercy for Nina and all the suffering and helpless beings in this world.

I recall hearing from some clerical pundit when I was a child that our purpose here on earth is to know and serve God, and we do that so we may spend eternity with that same God, sitting at His right hand, mind you. A boring prospect, I thought at the time, especially when I took a gander at all the holy people who were urging me to be good and thought that was who my neighbors would be. But during the days and nights of Willowbrook, days and nights of plotting and planning, of awful despair and uproarious laughter, I found people I'd be happy to spend eternity with.

Part II

Stages

During the late seventies and early eighties, I was sometimes very conscious of the galloping years and my lack of any real achievement, the general wastefulness of my life. I worked for a bit as a bartender at various places around town, and I wasn't too happy at my sudden drop in social prestige. I put too much value in that sort of thing, and it never occurred to me that I was important to my children, Siobhan, Malachy, Nina, Conor, and Cormac. I didn't realize that my opinions mattered to them, that they listened to and acted on what I said, and, as was the case on several occasions, they were hurt by my thoughtless comments.

While I was off gallivanting, Diana was creating a magical place for the children. She made puppets and a puppet stage, and had them painting and drawing and doing minidramas of their own creation. I was nearly always too tired or too hung over to participate, though.

Designing and building the puppet stage got Diana interested in doing more woodworking, so she took a course in basic carpentry. In no time at all she teamed up with a friend, Sheila Brown, to operate their own company, Dovetail, which specialized in cabinetmaking and kitchen design. As they ran up against so much discrimination against women in all of the construction trades, they eventually helped found Women in the Trades to combat the more egregious doings of the he-men on the job.

My work hours were highly irregular, and I had a lot more downtime than Diana, so I essentially became the house husband for a while, doing the shopping and cooking, and threatening the children with "Just you wait till your mother comes home!" It took a long time for it to dawn on me that this new occupation of Diana's was not a temporary hobby, that she was quite serious about earning her income this way. For all my supposed enlightenment and liberalism, I didn't see such philosophies applied to people in my own family. I was a great advocate of civil rights, human rights, and women's rights, and more latterly, gay rights and rights for the handicapped, but 'twas hard to live the spirit of those rights.

Looking after a house and making beds and cooking are great services to other human beings, but I found myself going bonkers with the seeming triviality of it all, though I couldn't think of what I would be doing if I wasn't doing just that.

Still, on many occasions, I was proud of Diana for breaking new ground for women in the trades. When we went out, there was always great curiosity about Diana's doings as a cabinet-maker. And yet, because she was a woman, many people would assume that her prices would be very cheap, and they'd hem and haw when they got the quotes for a job.

Although the business prospered, and Dovetail had its own shop with five employees (mostly male) the profit margin was low. It was a constant struggle just to make the payroll, and eventually they had to close the business. Not alone a business, but a social experiment that lasted for more than fifteen years thus came to an end, and there remain many finely built kitchens and studies and cabinets and bunk beds in New York because of Diana and her partner, Sheila Brown.

I **have never myself been much for manual labor of any sort.**
When I was fermenting as a child in Ireland, my destiny was to
be handed a pick and shovel and told to go dig something
somewhere. In those days, you were lucky if you could get any
kind of a job at all, even digging, and men went to their graves
grateful for the chance to break both stones and their health
with their labor. Not for me! The thought of relentless shoveling for all the days of my life struck fear into my heart.

The great ambition of every family in our milieu was that
the children should grow up and get a job that required a collar and tie (in those days, the collars were detachable), and was
indoors, away from the elements. The mothers talked in awed
tones of somebody's cousin's son having gotten a "clarke's" job
in a solicitor's office, or somebody else's daughter who was
now working in a bank. There was a particular reverence for
those who worked in banks, as if being in close proximity to
Mammon itself bestowed power and prosperity.

I didn't know what I wanted to be, but road digging, paper
shuffling, and money counting were not on my list of ambi-
tions. Policeman or fireman didn't appeal to me. A cowboy's
life seemed pretty good, as cows can be bossed around, and
that was important to Irish people. Adults would often talk
with great respect of a boss who had twenty or thirty men
under him. So, maybe I could boss hundreds of cows around
and boast about that.

Being a pirate was out of the question, as I was told they
went out with the sailing ships. I briefly considered being a
ship's captain, but that meant learning navigation, and they
told me I was no good at numbers. My daydreams usually
ended with a decision that when I grew up, I'd just go to
America, for there you can do anything, and none of it
involved digging roads or lifting heavy things.

Once I found it, the acting biz always seemed the thing for me. Having got my first job by just walking in and asking, it seemed like a cinch. People consider it an art, and I suppose making people laugh and cry and convincing them that you are someone else is some kind of art, but I was never convinced I was doing anything artistic. I think it was Spencer Tracy who said, "Hit your mark, look 'em in the eye, and speak the truth. That's all there is to acting." That's how it's always felt to me. Sometimes I was good in a part, and sometimes not, as I'd no training and no patience with the process of rehearsal. As for technique, that was to learn the lines. The one lesson I took away from all those years is, "Listen to the other actor! Just listen!"

Most actors are like nomads wandering the desert, hoping for, and sometimes coming across, an oasis in the form of a great play or screenplay. Even when successful, the actor is usually doomed to performing in a pile of horse dung, which in public he has to say is gold and that the stink is in the nose of the beholder.

When I decided to work exclusively at the acting trade, it didn't occur to me that being a full-time actor meant having a lot of time on my hands and very little acting to do, lots of high hopes which were forthwith dashed. Thousands are flocking to drama schools and graduating with BFAs and then heading to graduate school for the MFAs, getting ready for the big break with singing lessons, dancing lessons, dialect lessons, and coaches and classes of all kinds. It's good we have enough restaurants in Hollywood and New York to employ all those who would be thespians.

Of course, I shouldn't be talking. In my heart there was the hope that I would land a nice, cushy part on a series, which would make me financially comfortable for life. Instead, I ended up in bit parts in movies and readings Off Broadway, and being tempted to take a two-hundred-dollar-a-week stage job in

Alabama, where I'd have to pay room and board. But because it was work, I'd think about it, and even be flattered at the offer.

I don't know how many auditions I've endured over the years, for television, plays, movies, voice-overs, and commercials. I auditioned for commercials for products I wouldn't give to rats: soft drinks, junk food, and junk food franchises among them. I forbade my children to partake of these sugared and salted poisons, but quite willingly tried to sell them to yours.

The audition procedure goes like this: You pick up a portion of the script, the part pertaining to yourself (for some reason called "sides"), and study same, and then appear at the appointed hour to strut your stuff. Sometimes it is only a matter of putting you on videotape, but generally you get to read for the casting director, director, and producer.

I firmly believe that desperation gives rise to a specific body odor, which casting directors are expert at detecting, and if you are desperate, you are bolloxed before ever you open your mouth at an audition.

Sitting in a crowded casting office one day, I got into a conversation with a fellow who was very nervous and insisted on telling me he hadn't worked in over a year, and that he was going to tell the prospective employers he really needed the job, and what did I think of this tactic. Of course, I thought he was just being funny in a desperate sort of way, and I said I was going to do the same thing. He looked stricken, until I assured him I was jesting.

They give you jobs if they think you can do them, and if you stand to make money for them, so your need has nothing to do with their wants. However, they do rely on the passion and dedication of the thousands of men and women who want to be in the acting profession. Actors will work as extras, do bit parts in so-called showcases for no money; they will do staged readings gratis, and pay enormous sums for photographs and résumés and union initiation fees. They trudge from interview to audition to cattle call, early in the A.M., and sing and dance and perform many times just to see the show blasted by a critic, and down it goes, so the actor is out of work again. But, like those round-bottomed rubber clowns, they take the punches and come bounding up, again and again and again.

Unlike the computer business, the construction business, or investment banking, there is never a glut of employment in show business. For every job, there are at least forty actors, male, female, and child, who can qualify. Just to be in the acting trade calls for forbearance, the ability to take humiliation in stride, and to have full knowledge that success is rare, and that rejection is the largest part of auditioning.

I was sent to audition in Los Angeles for a part in a new series. As they say in the business, it wasn't much of a stretch, as the character fit me perfectly. He was a white-haired, partly Irish priest, with twinkly blue eyes and a sense of humor. I studied the sides, memorized my part, and turned up on time. I heard a voice asking was there a "Malachee McCord heah," and, despite the mangling of my name, I raised a hand. A young female, who looked all of fourteen, bade me follow her.

We entered a room, which contained no other human beings. I probably had a quizzical look when she announced I would be reading with her. But, fine. I began reading my lines.

"What, what is this?" sez she.

"What is what?" sez I.

"What is this dialect you're using?"

"This is not a dialect," sez I. "This is an accent."

"What is the difference?" sez she.

"An accent denotes a nationality and a dialect is a regional form of speech," sez I.

"Whatever it is, it just won't do!"

"The script says the priest is Irish," sez I.

"Not that kind of Irish," sez she.

I had to leave, lest I lose control and strangle this nitwit.

As often as not, though, even when you get the job, you can't be sure. In 1978, William Friedkin saw me for his production of *The Brinks Job*, based on an actual robbery. The part I read for was the Boston police lieutenant who was in on the heist. The audition was in Boston, and Friedkin told me to go back to New York and get fitted for two uniforms and two suits, which I did.

The following Monday, my agent called me and said she had bad news. The part of the detective had been written out, deleted from the story. I'd agreed to fifteen thousand dollars for the job and, as always, was at that moment teetering on the edge of penury. The agent said she was very sorry, etc.

"Ask them for the money," I told her.

She replied that there was no real contract and we had no standing.

"So you aren't going to do anything about this?"

"There's no point," she said.

"Well, I'm going to," sez I.

What she didn't know and I did was that under the Screen Actors Guild and Producers Agreement, a fitting for costumes is considered a contract. A phone call to Ralph Serpe, the producer, caused a bit of consternation. Ralph was already a bit over budget, and this straw would do the camel in. He offered me a smaller part at a much smaller salary. I carefully hung up the phone. It rang, and my man Ralph said we'd been cut off. I explained to him that I'd hung up on him because of his insulting offer, and that if he was prepared to be straightforward about how he was going to pay me fifteen thousand dollars, I'd listen. He had no ideas, but eventually he accepted mine. I got my fifteen thousand for four days' work, and had an extra week or so with Peter Falk, Paul Sorvino, Peter Boyle, and a very congenial cast, plus I got to meet the real characters who pulled off the real Brinks job.

There is always an odd happening on every film. On this one, it occurred while filming a scene that involved about seven hundred extras, dozens of production assistants and assistant directors, plus half of the Boston Police Department. It was very costly, to say the least, and great care had to be taken to make sure all went well. It was supposedly the day the Brinks gang were being brought to the court for indictment, and the rented local crowd entered into the spirit of the thing with great energy.

Many people thought of the Brinks lads as local heroes, with the exception of one Specs O'Keefe, who turned state's evidence and spilt all the favas to the authorities. The actor playing Specs, Warren Oates, was hit, kicked, and pummeled by the crowd of extras, who forgot that this was simply a movie. All of it got on film. Friedkin didn't mind, as that kind of realism didn't do any harm, even without the Stanislavsky method.

Later, when the postproduction work was being done, the film editors were invaded by a group of armed men in masks, demanding the negative of the big scene. Off they went, and shortly thereafter a ransom note demanding a million dollars or so arrived. But the original negative had already been flown out to California and placed in a vault in Universal Studios, so all the thieves got for their trouble was a working negative hardly worth the film it was printed on.

Some people have written that the stage has been the fulcrum for social change. I have come out of the theatre at times delighted, moved, or thoughtful, but never have the emotings of some thespian or the urgings of any playwright sent me out determined to overthrow the government or take to the streets. I would be astonished if anything I did onstage has had anything but a temporary and immediate effect on an audience.

I had my first chance to stir the masses of Broadway in 1982, in a play by Bill Davis, *Mass Appeal*. It had been a roaring success in an Off Broadway production featuring Eric Roberts and Milo O'Shea, directed by the caring and beautiful Geraldine Fitzgerald, and a decision was made to move it to Broadway.

Before that could happen, Eric Roberts had an auto accident that nearly ended his young life, and everything was postponed pending his recovery. The cast of two was already set, but they did need understudies, and I auditioned and got the job of understudy to Milo O'Shea. The production moved to the city of Boston for tryouts, and previews were received very well by all and sundry despite the fact that Eric Roberts was still getting used to new teeth, his broken jaw and fingers were still healing, and his medication caused spasmodic gestures that startled anyone conversing with him.

The opening night was fraught with the usual anxiety and tensions, but went quite well, and the reviews were excellent. However, at the modest opening night party, Roberts announced his decision to quit, as he was not happy with the play's production. Panic thundered down on everybody, as now the understudy, a young actor named Charley Lang, had to go on. He acquitted himself well, despite the lack of proper rehearsal. But the Angel of Doom had not finished yet. The next morning Charley called the stage manager and managed to croak out what was immediately obvious: He had laryngitis, strep throat, and a high fever, and could neither rehearse nor, indeed, go on that night.

Again, panic, and I was warned that with all these disasters, I'd better be prepared to go on at a moment's notice. But I knew Milo O'Shea better than they did. The only thing that would keep that man off the stage would be death itself. Nonetheless, I kept my counsel and learned my lines.

The show must go on, and the show did go on in Boston, because the playwright, Bill Davis, was summoned, and, he played the seminarian to great applause and acclaim. Charley Lang recovered, and took over the part again, whilst a frantic search went on in New York for a star replacement. A lad named Michael O'Keefe was finally cast, and *Mass Appeal* opened to quite good reviews in New York. Though it was not a resounding commercial success, I did get to go on a few times, thus a McCourt played Broadway, not something they would have predicted back in Limerick.

One night in 1976, after an hilarious evening of recounting our lives at a family gathering, the brother Frank said he and I should put this stuff on stage. I wasn't at all sure anyone would be interested, but I contacted a lad I knew who had a pub with a theatre upstairs. He said we could use it for free, and in 1977, without script, costumes, or lights, we proceeded to get up on that little stage and tell our stories.

The show was raw, bordering on the amateurish, but people liked it. Pat O'Haire, of the *Daily News*, was the only reviewer to come when we first performed it, and she was kind. She wrote that it was better than going to a psychiatrist.

The mother, who was not yet so sick that she couldn't come see the show, didn't respond so positively. She was embarrassed when she saw it, and the night she was there stood up in the audience and pronounced it all a pack of lies. I invited her to come on stage and tell her side of the story.

"I will not!" sez she. "I wouldn't be seen on stage with the likes of ye."

In 1984, a friend mentioned that an acquaintance of his, Mike Houlihan, was looking for a project to produce. He'd heard about *A Couple of Blaguards*, and he wanted to discuss it with us. He told us he was most anxious to produce it and asked us to sign a contract giving him residual rights for fifteen years. With nothing to guide us but our own lack of experience, we blithely signed up, with hardly a cursory look at the documents.

It's a cliché to say that one should have legal advice before signing anything, but let me present to the reader a sage bit of advice: Get legal advice before signing anything. As a result of the McCourt success in a variety of fields, Mr. Houlihan has done quite well with *A Couple of Blaguards*.

The show opened at Art D'Lugoff's famous Village Gate in time for Green Ghetto Time, St. Patrick's Day. Some critics liked it, and others thought it a heap of shite, as the more basic Irish say, but it had a solid three-week run. Our doughty producer then suggested a run of a couple of weeks in Chicago, followed by another couple in San Francisco.

The show was set to open in the Windy City on July 5, 1984, as Frank would be free from his teaching job for the summer. We were booked to play at a strange venue called Cross Currents, an alternative gathering place for Republican transvestites, gay cowboys, communists for Jesus, anarchist accountants, and any other fringe group desirous of renting space.

If any show arrived in Chicago with less notice than ours, I've yet to read about it. The most significant public relations event we attended was a radio show which was not only on a station that couldn't be heard through a drying shirt, but was not scheduled to air until a week after we left Chicago.

Our producer also tried to arrange a press conference in City Hall to kick off the show. Not having had any experience in matters of this kind, we accepted his assurance that the public relations office had set it up and that it would be well attended by the press. Of course, the press didn't turn up, and why should they? A dinky little show opening up in a run-down cabaret is not headline stuff. Moreover, the Chicago City Council was having one of its famous wars, and the press were having a grand time with all of that stuff. Our producer chased reporters up and down the halls, trying to persuade them to talk to us, but they soon escaped to the serene lunacy of the council chamber.

But soon after we opened, Sheila Heneghan and Michael Cullen, a couple of astute and intelligent producers from Chicago, took over. For a miserable stipend, they immediately put a professional stamp on the show. They got the important reviewers to come, and there were so many interviews and appearances on radio and television that with all the comings and goings, we found ourselves shaking hands with one another.

Heneghan and Cullen made the show a hit. The *Chicago Tribune* called the show "the sleeper of the year," but you know you've really arrived when somebody satirizes you. There was a review at another cabaret in town, and it soon included a sketch called "A Couple of Blowhards."

With Cullen and Heneghan at the controls, we triumphed over ineptitude, anonymity, and chicanery, and we were ensconced. Our producer returned to New York and had thousands of dollars delivered to him every week. Of course, he decided to extend our run. When the school holidays ended, we had to hire a lad named Sean Tracy to fill in for Frank on Fridays. The brother would do his five shows each weekend, and return to New York on Sunday night, by way of People Express. (People Express charged seventy-nine dollars one way during the day, and fifty-nine dollars after nine at night, so the big airlines set about stamping out this pesky little upstart, which did not take them long.)

I was doing *Search for Tomorrow* at the time, and there was many a day I commuted back and forth from New York after doing a segment. I also managed to squeeze in a week on a movie called *Brewster's Millions* with Richard Pryor, before his explosive bout with crack cocaine.

The little theatre we were in held about two hundred people, and the management there ran the sort of amateur operation audiences might find charming—box office phones left unanswered, floors unswept, stage lights left dark—but actors do not. The management was so bad that our now well-to-do producer stopped paying rent at one point and they never noticed, because they were making so much money selling drinks to our audiences.

It was a languorous summer, and Lake Michigan was convenient to our lodgings for the swim. While we were there, the brother Frank, now divorced from his first wife, told me he was getting married again, to a lady named Cheryl, from one of the Carolinas. In general, the McCourt Family modus operandi is to object to whatever any other family member is about to embark on. Frank, being the senior member, must have sensed the objection trembling just below my tonsils, because he immediately barked that he hadn't told me in an attempt to solicit advice. He had informed me, he said, as he didn't want me to be puzzled when I received an invitation to the wedding.

I gulped and nodded and did not pound him on the head for marrying a woman he barely knew. On the other hand, he did not pound me on the head when I got involved in deals and doings far in excess of this particular escapade. A swift flight to New York City, practically between shows, got the ceremony performed, again by the McCourts' favorite rabbi, Bruce Goldman.

The mother Angela would have been disappointed once more, as not alone was her offspring divorced, but he was marrying another Protestant in a ceremony performed by a rabbi. During the ceremony, I could almost hear her: "Not a wan of yer has married a nice Irish Catholic girl. I don't know why ye can't stick with yer own."

With the exception of the wedding and the occasional bit of filming, I stayed put in Chicago throughout the run of the show. There are times in life when a blankness envelopes the soul, and all existence seems like a dreary trudge, and that is what happened to me toward the end of 1984. It wasn't exactly a depression, or perhaps it was, as I was still drinking and wandering around Chicago's bar scene, and though alcohol might seem to give you a lift, it is ultimately a swift elevator heading in the other direction.

I remember the eighties to be a letdown decade as a whole, and many of us were burned out on the marching, the demonstrating, and the sit-ins demanding that our government do the right thing. A deadness enveloped the times, and the soul of America was squeezed dry of any vestige of morality or compassion for those in need. It was a decade best exemplified by the Reagan administration's designation of ketchup as a vegetable in the school lunches of the poor children of America. I couldn't understand how any human being, let alone a president or his appointees, could face a nation and say that a sauce laden with sugar could be called a "vegetable." Nor could I understand why the American people couldn't see through the fake charm and pseudosincerity of this untalented actor and his Lady Macbeth.

The mean, muckworming rapacity of the Reagan years has imprinted itself on my memory. Every move of that grinning administration was designed to enrich some greedy supporter, while at the same time depriving people on the lower rungs of society of nutrition, education, and opportunity.

My mother had always spoken glowingly of America when we were children in Limerick. A bit of time and a bit of distance had dulled her memory of the death of her only girl child there, and of the hardships she endured whilst chasing after my father in the saloons of New York for food money. She told us instead of the wonders of the great American city, the rumble of the subways, the swarming crowds and how they were dressed in bright colors, unlike the drab garb of the Irish, and of how the restaurants, whether exotic or ordinary, never ran out of food.

She knew a bit about American history, too, and that the government was of the people, by the people, and for the people. She told us that in America, the government cared for the poor, unlike the Irish government, which served the privileged. There was never any doubt in her noggin that America would do the right thing all the time.

"Wasn't Franklin Delano Roosevelt the greatest man who ever lived? Didn't he save the workingman and his family from starvation and from tyranny?" said my mother, and the sooner we all got back to America, she knew, the better off we would be. And wasn't I convinced? That ideal of the United States and its government has never completely deserted me. In fact, I still believe in the integrity of most men and women in public office, especially if they are not of the governing party of the minute. I'd met some of them—Robert Kennedy, before he ran for the presidency, and that grand, decent man, Eugene McCarthy, who restored honor to that sullied name.

I'd also met Jimmy Carter, as decent a man as ever put a shoe on his foot. In a country used to horrendous waste and misuse of the world's natural resources, there was a lot of small-minded sneering and scoffing at Carter's policies for conserving oil and utilizing solar power, but he was a man of good heart and great foresight. I did work in his campaign, and it was painful to see the headlong flight of frightened and greedy people from the Democratic party. Those people went and elected Ronald Reagan, Actor, a man I consider the greatest nincompoop king of all the thieves and robbers masquerading as Americans this country has ever seen.

I met Carter at a breakfast shortly after the Democrats had nominated him, and he startled me by asking what I thought the most pressing issues of the day were. I wasn't prepared for a presidential candidate querying me about affairs of state, so I stammered something about the need for taking care of the nation's physically and mentally handicapped citizens. I must say, not only did the man not yawn when I broached the subject he proceeded to ask me a few questions, which were quite intelligent. I was a loyal follower of President Carter from that day on.

Ah, yes, I believed there are good men, and still do, but there are times I want to go to bed and pull the bedclothes over the head and shut out the horrors of the world. Chicago in 1984 was one of them. This was brought to a halt by the producer, when he decided to close the show in January. As livings have to be earned, rent paid, and food supplies secured, I couldn't stay in bed forever, despite the dark despair. As it turned out, the producer closed the show because he had secret plans to make a huge killing by doing a marathon series of the show a couple of months thereafter. Foolishly, when it was finally revealed to us, we went along with this plan, and packed thousands of South Chicagoans into a theatre to enrich this man.

Shaking off the paralyzing malaise, and looking to the future, I discussed with the brother Frank the idea of taking *A Couple of Blaguards* to Ireland in the summer of '85 and, as the man was agreeable, I hopped on a flight over to Limerick. I had a swift consultation with the management of a little theatre called the Bell Table, and forged a agreement to produce the play there. Hardly any money was involved, as they didn't expect big audiences for a no-name show with practically unknown performers, Frank and myself. Moreover, most Limerick people of any means whatsoever took themselves to the seaside in the summer, if they had any sense at all.

July came, and Frank proceeded ahead of me, to take care of the usual preparatory stuff. We had decided that this journey would also be an opportunity to bring the mother's remains back to the city that spawned her. The tin can containing Angela's ashes had been knocking about our various apartments, and at that moment they were in the custody of Frank and his new wife, Cheryl. The deal was that Frank would bring them to Limerick for a bit of a ceremony, taking advantage of our presence there. But Frank forgot the ashes, and rang me in a panic from Limerick to go and get them from Cheryl. Diana went, and had to root in a heap of possessions in the cellar to retrieve them.

As we had a paucity of funds, I booked Diana and self on a chartered flight to Shannon Airport, at a fare considerably below that charged by the regular airline, Aer Lingus. But, as one finds out, there is a considerable difference between cheap and inexpensive, and charter flights are nearly always cheap. This particular one turned out to be the flying equivalent of an ass and cart.

Diana and self arrived at Kennedy Airport at 7:00 for a departure at 9:30 P.M. We went through the usual formalities of swearing we had packed our own bags, and seeing as we had no bombs or guns or explosives, the only thing we had to explain was the sealed tin can, containing the ashes of the mother. We had opened it before, as Diana had suggested we should keep a few remnants, which we placed amongst a collection of sea stones we kept in our apartment.

I popped the lid and satisfied the curious security folk that there was nothing but little bones and ashes nestled in the can, and then we waited and waited and waited for the boarding and departure hour, which came and, of course, went. The clock hands hit midnight, and officials of this ragged charter company told lie after lie about what was causing the delay. A large family group on their way to a wedding began dipping into their duty-free liquor, and a pretty good sing-along erupted.

By 2:30 A.M., the airport was deserted, and not a drop of refreshment or food was available. One of the wedding party erupted into a mighty roar of frustration, and let loose an eloquent torrent of profanity and blasphemy, the best I'd heard in many a year. Official reaction was not as appreciative as mine, and he was arrested by the airport police, which I suppose was intended as a warning to anyone else who chose to complain.

Finally, at 5:00 A.M. we got on board the plane, buckled up, and departed. There wasn't much to eat or drink, so we pretty much settled down to sleep. The wedding party were sunk in a postalcoholic slough of despond, hung over and dispirited at having lost one of their stalwarts. Diana got to chatting with a retired policeman, a former member of the notorious NYPD Red Squad, which spied on and photographed people who demonstrated against the war, or anything else for that matter. It turned out that we had been to many of the same demonstrations, on opposite sides, of course.

As happens on aeroplanes, the hum of the engines stills talk and people begin to nod off. But a nagging and ragged whistling sound kept intruding on the hearing and causing restlessness. It got louder and the anxiety rose further at the sight of a stewardess, with a stricken look on her visage, galloping toward the cockpit. In a swift about-face, a man with the gold stripes on the sleeve came galloping from said cockpit, headed in the opposite direction, followed by the stewardess.

Nearly every conversation at a time like that begins with "I wonder what . . . ," and this time was no exception. Instead of wondering, I hefted myself out of the seat and ambled as nonchalantly as possible toward the rear of the plane. There I observed the striped one energetically stuffing newspapers into the area 'round the edges of one of the aircraft doors. The people sitting in the area looked terror-stricken, with fingers stuck in their ears to keep out the piercing, high-decibel whistle coming from the leaking door.

The gold-striped one got enough newspapers to get the noise down to a reasonable level, and said, "That should do it 'til we get to Shannon." He returned to the cockpit, and then the voice of a man identifying himself as the captain said the air leak in the door was nothing to worry about, as it was like opening the window of your car a crack while traveling at sixty miles per hour, just noisier.

We settled down again to sleep, with the engine's hum and the now lesser whistle in the background, until the whistle once again soared in decibel level. Once again, the stewardess galloping up the aisle. Once again the striped officer galloping down the aisle. He returned to the cockpit and the captain once again asked for our attention. "Nothing to worry about," sez he, "but we are about to turn 'round and go back to New York, due to the 'annoyance' of the cracked door. For safety's sake, we are going to descend to eight thousand feet."

There was a flurry of moans and groans, and people began kissing scapulars and holy pictures, throwing holy water around, and rattling rosary beads, and rocking to and fro. As the plane nosed down toward the Atlantic Ocean and made a very wide U-turn, my own thoughts became more reflective than fearful. It occurred to me that I ought to tell Diana that I loved her, despite all the rotten things I'd done. It also occurred to me to order some whiskey, but I didn't, and for the first time in years, I was facing a crisis without alcohol or denial.

I thought and decided that if I start telling the woman I love of my love for her, it will only make the crisis seem more deadly, so I didn't. I just held her hand, and, like many frightened animals, I simply drifted off to sleep, musing at the thought that I'd always hoped to die in my sleep. Drifting in and out of my slumbers, I remembered stories of aircraft doors being blown open and passengers being sucked into the atmosphere, never to be seen again, and how appropriate the hymn "Nearer, My God, to Thee" was at that altitude.

At some point, I began wondering about this God fellow, if there was such a thing anywhere. "If there is a deity," sez I to myself, "I'm going to feel pretty stupid when I have to face It, as I've been going about the world bellowing, 'I'm an atheist, thank God.'"

The more religious told me I'd be screaming for a priest on my deathbed, and now here I was, a thousand miles from my bed, and probably just as far from a priest, and not unduly upset about the lack of either. I'd more pressing items on my agenda, such as wishing I'd been a better husband. I would have loved to have another chance to hug my children, and to tell my brothers I wished I'd not been so judgmental of their lives and doings. Then I thought that I'd been too judgmental of most everyone I'd met, but upon further reflection, I was delighted to have thumped a reasonable number of arseholes who infect our world.

There are writings on near-death experiences that say your whole life flashes before you at the time, whereas I found it took a while to get the whole bloody thing in order and proper sequence.

The ocean, which had looked majestic yet friendly from thirty thousand feet, looked ominous and threatening now that the plane was practically surfing the waves at eight thousand feet. I wondered whether we would go straight down to the bottom or float for a while, and then I snoozed again, calmly wishing for some resolution to the situation.

Our captain kept reassuring us that everything was under

control, but the stewardesses were obviously scared. I found out later that most of them were stewardesses part-time and not particularly well trained for emergencies.

We did get back to Kennedy, in a couple of hours. The passengers were put up at a nearby motel at the charter company's expense, and after a few hours we were summoned back to the airport to begin the journey anew. Without actually saying so, the people from the company implied that we had a different plane, and that all was well. About twenty-five passengers opted to stay off and take Aer Lingus later that evening, not wanting to risk their lives to save a few hundred dollars. Being fairly close to penury at the mo', I didn't feel we had that option, and away we went into the tame blue yonder, different seating this time, and much more room because of the departed passengers.

Out of curiosity, I went back to our previously assigned seats, and, after digging in the magazine pocket, I found a periodical I'd left there from the other flight, which, of course, proved we were on the same plane again. Diana had already noted that, but I had pooh-poohed the observation, on the grounds that the folks who run this company have no reason to lie and, furthermore, I was sure the crew were as anxious to keep on living as we were. As we were airborne, it was too late to do anything about this deception, and so we settled down again to the droning boredom of temporary life inside a mutated toothpaste tube.

My intermittent snoozes were interrupted by a familiar sound, a high-pitched, high-decibel whistle impinging itself once again on my consciousness. Of course, I thought I was dreaming a bad dream, reliving a nightmare, but no, 'twas a reality. Once more we had the scenario of the stewardess galloping toward the cockpit, followed by a gold-braided man galloping toward the rear, followed by the reassuring announcement from the captain. The difference this time was that we had already proceeded beyond the point of no return and had no option but to keep flying to the Emerald Isle, or perdition, as the case may be.

To divert people's attention from the possibility of their immediate demise, I asked if anybody would like to meet my mother. Even at death's door, most people are polite, particularly when it comes to mothers and meeting them, so I got a positive response. I reached into the overhead bin and pulled out the can that had once held beans and now held my mother, marked "Ashes of Angela McCourt," and held it up for all to see. The reaction was not a happy one, as all sorts of superstitions and fears came to the fore, and I was requested forcefully to put it away, as it was bad luck. Thus ended my attempts to divert the folks.

We descended again to skimming the waves, and the unfortunate people in the vicinity of the screaming door had to put pillows and blankets to their ears to keep out the dreadful sounds. For another three hours we trundled in contained misery to Shannon Airport in Ireland, and landed with a rousing cheer of relief from the passengers. The good brother Frank was there with transportation to get our weary bodies to the bed on terra firma.

A Couple of Blaguards is a relatively simple show to do. All that's needed on stage are a couple of chairs, a table, and a bureau or chest of drawers that can be used as a bar, a pulpit, or a judge's bench. For ourselves, we had a couple of glasses, and some women's wigs and a couple of scarves, a stick or two or three, and hats for the various characters. It's not exactly an Andrew Lloyd Webber musical. The lighting is simple, and movement is minimal, as the play depends on words rather than gross action. There is only the matter of memorizing what Frank and myself wrote. At the Bell Table in Limerick, on that Fourth of July 1985, we had a quick run-through in the afternoon, and that was it.

At one time in New York, the mother had told a friend of ours, Sheila McKenna, that if she'd ever had the chance in life, she would have loved to have been an actress. This was news to us, her sons. She'd never said she would like to be anything in life. It's likely she thought she would have been laughed at for daring to think she could achieve anything, let alone being an actress. But when we talked about it, of course, we remembered her romanticism, the love songs she sang, her ability to recount nearly every scene in every film she'd ever seen, and how she talked about going to see live theatre when she was a child.

The Bell Table was an old converted cinema, formerly known as the Coliseum, where Angela, whenever she could get the few pennies together, would go to weep over the blighted love lives of the rich and infamous, as filmed by Hollywood's cinematic storytellers. She never lost her almost childish wonder at the glamour of it all, and there were times we were glad we didn't get to see a movie, because she had recounted it so well. Keeping that in mind, I placed the bean can containing her ashes in the chest of drawers on the stage.

The reviews of the show locally and nationally were quite good, but the houses were quite small, as expected. One upstanding citizen, Charlie St. George, an old family friend, took grave offense at our depictions of some of the illustrious sons and daughters of that ancient city, and never spoke to us again. He also went to the *Limerick Leader*, a local paper, and demanded they print his review. In it, he said we had dragged the good name of Limerick through

the mud and slandered good and respectable people, and—
my favorite quote of all—"used the stage to dip into verbal
pollution."

I thought his use of the language was wonderful, and
thought the fact that the local paper would print a review by
an ordinary citizen was pretty admirable, as well. Needless
to say, we posted that review outside the theatre, where it
drew considerable comment and maybe got us some customers.

On the closing night of the show, Frank and myself spoke
the customary words of thanks to the audience, and said that
our mother thanked them, too. We brought out the bean can
with the ashes and told the audience whom it contained. As we
were going to scatter them on the morrow in the family grave-
yard, we asked them, would they please rise and give Angela
her first-ever standing ovation, a request they enthusiastically
honored. That applause was surely heard in heaven, to the
amazement of God and his Right-Handers.

I'm not sure what Thomas Wolfe meant by his oft uttered cliché,
"You can't go home again." Obviously, you can go home again,
today, tomorrow, or tonight. But does it mean going home as you
did after school, after work, or after playing a rugby game? For
me, this going home was supposed to be a somewhat triumphant
return of the boys, who'd done well despite the poverty, the social
humiliations, and the discouraging words heaped on us in Limerick. It wasn't that we'd become big stars like other men from
Limerick, such as Richard Harris in Hollywood, or Terry Wogan,
who became a famous broadcaster in Britain, but I'd done well on
stage, television, and radio, and Frank was a highly respected
teacher and well regarded by journalists and writers in New York.

Dreams of returning in some kind of a glory halo worked their way into the wounded soul. I had left Limerick an anonymous, embittered twenty-year-old in the clothes of a dead uncle, Pa Keating, and I'd been back a few times, but nobody had noticed, and that didn't suit me at all. Now I was going back after achieving a bit of fame as an actor, so I hoped the requisite respect would be paid.

In retrospect, I think my agenda was that there would be fulsome apologies for the way they had treated us as children, and that the entire city would confer honors on us as reparation.

My agenda was not followed.

People said it was a good show and well done, congratulations, but nobody mentioned the past. Nobody said how amazing it was that we had overcome the difficulties in our lives and gotten over all the obstacles that had been placed in our paths. On the surface, I remained the genial Malachy, and I had the strange illusion that I was enjoying myself, because of the fact that there was never an inappropriate time for drinking. But I drank angrily and managed to provoke stupid arguments with Frank and Diana, which allowed me to stomp off and spend time in the pubs with strangers.

The Irish pubs have always held a fascination for me. They were the places that claimed my father's time and money, and those of his friends. There were times we formed little posses to go out and find him; I'd find myself amongst men who smelled of sweat, porter, and tobacco, all standing at the bar. They seemed jolly and teasing, and sometimes they gave us a ha'penny or two. There was warmth and light and camaraderie there, and song and talk and laughter.

I was still looking for my father, I suppose, and somehow believed that if I stayed long enough in the pub, I would find him. In the process, of course, I would lose myself.

But the pubs were not the places of my memories. I wasn't straining my neck looking up at enormous men with gaunt, stubbled faces and teeth scattered like little yellowing tombstones in dark mouths. Now, as I stood at the bar amongst them, most communication was through imperceptible nods for refills or greetings to a newcomer, and conversation was muted, lest the sacredness of the atmosphere be polluted with idle chatter.

There was gloom and quiet madness in those places, and an implacable stolidity, holding fast against the incursions of joy and laughter. I never found the much-vaunted "salon saloon" of great and celebrated Irish wit and conversation, of learned discourse punctuated by quotations from the Greek, the Latin, and even Shakespeare. All I found were lonely gray men staring ahead and nursing a pint of porter, which gave them the right to sit or stand at the bar until it was time to shuffle off to a lonely and lonesome bed. There was no comfort for me in those places, as the fear grew in me that I would finish my days a graying specter of depressed humanity, encased in isolation, nodding to a barman for a refill.

At the same time, on the horizon, a dawn glow was just barely making its appearance, throwing soft light into some dark corners. Francis Thompson's poem "The Hound of Heaven" announced itself to me, and started on slippered feet its cadence.

> *I fled Him, down the nights and down the days;*
> *I fled Him, down the arches of the years;*
> *I fled him down the labyrinthine ways*
> *Of my own mind; and in the mist of tears*
> *I hid from Him, and under running laughter.*
> *Up vistaed hopes I sped;*

> *And shot, precipitated,*
> *Adown Titanic glooms of chasmèd fears, . . .*

The drink was no longer working for me as well as it once had. Instead of numbing the pain and raising the spirits, I was feeling like Falstaff in *Henry IV*, Part I, after the prince laid a pile of rhetoric on him: "There is a devil haunts thee in the likeness of an old fat man, a trunk of humors, that bolting hutch of beastliness, that swollen parcel of dropsies, that huge bombard of sack, that cloak bag of guts . . ."

In **Limerick, I could no longer successfully suppress feelings** I'd ignored my whole life; couldn't seem to clean out the rage and bitterness and depression that were clogging my soul and spirit. At intervals, the whiskey let me forget for a bit, but then I'd slide deeper into the slough, when sobriety returned.

It was this time in Limerick, too, that memories of being molested sexually came flooding back. I was about eleven when one of the priests told me he would test my ability to resist sin, and then he fiddled with me, all the time lecturing me on the punishments that awaited me in hell and the lashings with cat-o'-nine-tails I deserved.

Another priest, another time, took a jolly, friendly approach. "Let's see how much you weigh now, lad," he'd say, and lift me up, hugging me to him. "Hmmm. Nine stones is it? Nah, that can't be." And he'd adjust his grip and heft me again, up and down, and the whole time I could feel him pressing against me, and I knew he didn't give a fiddler's fart for how much I weighed. There were incidents with three different priests, and another man, a layman but a pillar of Limerick's church. I had never told anyone what happened, for it was obvious to me at the time that it was my own fault that these men were groping at me, rubbing themselves against me. I knew I must have done something to deserve this. Hadn't I been told I was a sinner already, hadn't they told me that just about every day at Leamy's and on Sundays in church?

Over the years I'd had vague memories, but it was something I didn't want to think about, and I made it a point not to. But now, with the whiskey no longer to be depended on, in the place where it had all happened, I couldn't ignore the feelings anymore, feelings about myself, and those men, and all the other adults who let these things happen. It did not help to know that these holy pedophiles were all dead now and couldn't be brought to justice or confronted.

You *can* go home again if you are prepared to face your memories and endure the remembrance of happiness fled and of people dead. You *can* go home again if you have a heart as big as Eternity to forgive all that was done to you, and to wish heaven's happiness to those who may have damaged you. You *can* go home again if you can embrace all who are still living there and don't know, or don't want to know, about the past and what place it has in your soul.

Before we left Limerick, there was a ceremony scattering Angela's ashes at her family burial ground, Mungret, which is a tenth-century abbey, now a national monument. I thought there must be nobility in our family, if the place of burial was the floor of a tenth-century abbey. I was slightly off in this department. It turned out that my ancestors were servants and scullery maids for the priests, and, as the old abbey was abandoned, my ancestors were allowed to bury their dead there.

Now that Ireland has come to its senses about preserving

these ancient edifices, we are no longer allowed to bring our dead to the old place. So, when we went to scatter the mother's ashes, we were confronted by a barred gate, but there was enough room between the arch and the top of the gate for us to clamber over into the old abbey.

Frank and his wife, myself and Diana, our son Cormac, who joined us, were all there, and a few old family friends from Limerick came, too, and we said some words about coming home and resting in peace. We scattered the ashes and then put the bean can in a niche in the wall, and then went off and had a celebration meal now that Angela was in the bosom of her mad and lunatic dysfunctional family, whose traits I inherited.

After Limerick, it was time to move the show on to Sligo— Yeats country. On my previous trip to Limerick, Sean Costelloe and myself had hied our way to the Hawke's Well Theatre in Sligo to discuss bringing *Blaguards* up there after our run in Limerick. After our discussion finished, and a few pints of porter were put down on top of the dinner, we set off to return to Limerick, a couple of hundred miles to the south of us, and promptly got lost.

The country road we were on was darkish, with no habitations in sight, but out of the darkness strode a big man wearing the requisite workingman's cap, the Wellington boots, and a big black coat tied with a piece of rope around his waist. I hailed him with an "Excuse, me, sir!" and he turned and slumped over to our car.

"What can I do for ye?" sez he.

"Could you tell me the way to Limerick?" sez I.

"I cannot," sez he.

"Thank you," sez I, and we turned to drive on.

"Hould on now," sez he. "I can tell ye the way to Clare-morris, and I hear tell there is a road from there to Galway, and they tell me there is a thoroughfare from there to the city in question, Limerick."

"That will do fine," sez I.

"Now," sez he, "first of all take this motorcar that ye and the other gentleman is sitting in and turn it around to face whence ye came, for if ye continue on this road, ye will be going around the world for sport, seeing as the Atlantic Ocean ahead of ye is the only thing between ye and America, and proceed down this road and turn left in a half mile, and that's the road that will take ye to Claremorris, and thence to Galway and farther, so they say." He was not going to commit himself. "Ye can't miss it. Good night to ye, and may God see ye safely to yeer destination."

It was worth the whole trip to Ireland to hear speech like that, and the result was a week performing in Sligo. Our audiences were small again, but enthusiastic, and the Hawke's Well is a state-of-the-art theatre, so it was a pleasure to perform there.

The twelfth of July in Northern Ireland is a big holiday, the day the Orange Orders celebrate the triumph of King William over the Catholic King James, who complained to the wife of one of his generals that his Irish troops had run away, and received the reply, "Your Majesty seems to have won the race." On that day there are endless parades with huge drums called lambegs, beaten 'til the drummers' hands run red with blood, defiant songs are sung, toasts are proposed, and much fun is made of the pope and popery.

Diana and self made our way to the border to go to Enniskillen to see one of these energetic spectacles. On the Republic side, we were stopped by an Irish policeman, who asked if we were going to the parade and, without saying so, made it clear that it was none of his damn business where we were going. He said he was sorry to inform us that there was a large unexploded device under the bridge and that we would have to make a five-mile detour, as the demolition people hadn't arrived yet.

"That's okay," sez I, "as I would rather live."

"You are right there, sir," sez he, and off he strolled, and off we went on our detour.

There were ninety bands in the parade in this little town, and the men played fifes and drums and displayed Battle of the Boyne banners. Watching the parade, women huddled under umbrellas and draped themselves in the Union Jack, and men dressed in the dark suit and bowler hat, and draped around their necks was the orange sash.

Diana asked a group of the huddled women if she could take their picture, and one said, "Yes, provided ye don't use it for anything."

We walked out to a field after the parade, where all had gathered in the now pelting rain to listen to politicians and clergymen talk of the horrors of unification with papist southern Ireland and the possible elimination of all things Protestant and royal.

It was cold and wet on that hillside, and we wended our way to a nice warm pub and sat and listened while the paraders made fun of the speakers and sang their songs.

"To the glorious and pious and immortal memory of King William III, who saved us from rogues and roguery, slaves and slavery, knaves and knavery, from brass money and wooden shoes, and whoever denies this toast, may he be

slammed, crammed, and jammed into the muzzle of the great
gun of Athlone, and that gun fired into the pope's belly, and
the pope into the devil's belly, and the devil into hell, and the
door locked and the key kept forever in an Orangeman's
pocket, and here's a fart for the bishop of Cork."

A few years later in that same town, at that same parade,
bombs were set off, killing a dozen or so men, women, and
children. In the name of Christianity.

When we returned to the United States, I felt wretched, both physically and mentally, so I went for a checkup. Having just returned from the land of stout, nicotine, and fried food, I was not a fine specimen of fit manhood, but the physician declared he could find nothing amiss in the Malachy McCourt corpus except that he felt I was drinking too much, smoking too much, and eating too much. I ought to stop smoking and drinking, and do something about my eating habits, he told me, and that some regular exercise wouldn't do me the slightest bit of harm.

I didn't see knocking off the drink as an option—perhaps I might cut down on it, and drink beer and wine instead, which to me amounted to total abstinence. And, certainly, a drink without the cigarette was out of the question, but I resolved to think deeply about exercise, as that in itself is quite strenuous.

Then the mind meandered in the direction of the damage caused by fattening foods. I had the temerity to ask the doctor about some appetite suppressants, which caused him to raise an eyebrow and suggest I apply some thought to regaining my health without recourse to chemicals, a variety of which were contributing to the mess in the first place. The remedy was in my own hands, he told me.

"And what might that be?" sez I to him.

He put his fingertips together and informed me that if I didn't leave his office and stop acting like the complete fool, he would double the fee. That got me through the exit in double-quick time.

I settled down to address the weight problem. Some people suggested fasting one day a week; others said eat only steak. One guy offered that injections of the urine of pregnant women had worked wonders for a friend of his. There were the "Dr. This" diet and the "Dr. That" diet, the protein diet and the grapefruit diet. There was the California diet, the Miami diet, and the Westchester diet. I didn't know if I was going to lose weight or take a tour of the country.

A good friend suggested that if I were of a mind to go to a program that dealt with overweight and other eating disorders, I might give a thought to going to a meeting of Overeaters Anonymous. I narrowed my options down to OA or Weight Watchers, and since OA was free and Weight Watchers wasn't, I hied myself to a meeting of OA.

I sat with a group made up almost entirely of women, of every age, and heard horrendous stories of anorexia, bulimia, starvation, bingeing, and other stuff I never knew went on in this prosperous society.

As I continued to go to meetings, I didn't hear too much talk about cooking or slimming foods or fast weight losses or good recipes for getting the fat off. People talked about their "relationships"—stories which made me groan, as nearly always it was a relationship with some destructive jerk—and they talked about cats dying, and dogs getting hit by cars, and not getting along with parents. They talked about everything except what I came for, controlling my appetite.

They did talk about food, but when they did, it made little sense to me. People talked about "stuffing down feelings with food," and I was too afraid of being scorned as an ignoramus to ask what they meant by that phrase. In the tradition I came from, you didn't show feelings because it was bad manners to go blubbering and caterwauling about life's upsets. So, I eventually learned, you got fat or became alcoholic, or went mad (which is the same thing), 'cos you couldn't talk about what ailed you.

On the walls at the meetings were charts listing the twelve steps of Alcoholics Anonymous. At OA meetings, people said just substitute food for alcohol when I read the steps, and I'd be on the trail to recovery. My eyes kept wandering back to the wall chart, which said: We admitted we were powerless over alcohol, and that our lives had become unmanageable. We came to believe that a Power greater than ourselves could restore us to sanity. And, we made a decision to turn our will and our lives over to the care of God *as we understood him*.

My reaction to any kind of religious evangelicalism or rigid dogma has always been to tell 'em to shove it up their jumpers. In Limerick, I'd watched some of the more upstanding citizens cheat poor people but give money to the church to ensure their salvation. I saw hypocrites making great shows of piety in church, beating their breasts—"crawthumpers," we called them. I'd seen the holy priests staggering drunk, at times driving fancy cars, and I'd been molested by some of them.

My rage had always prevented me from being able to distinguish between religion and spirituality. What was going on here, in this odd program about food abuse, was spirituality. I didn't want to have anything to do with the God business, I told them, and nobody was angry about that. They told me I'd be just fine, and someday I'd find a higher power, a god of my own understanding. It wasn't like anything I had experienced before. I had hove into a safe, nonjudgmental harbor.

I suppose those damned charts had an effect on me. It was impossible not to notice that the steps said that alcohol was the prime culprit in the destruction of people's lives. About a month after I began going to OA meetings, I copped to the fact that excessive drinking was indeed my problem, and that I'd better begin addressing the biz of booze.

So I stopped drinking, just like that—suddenly, precipitately, and perhaps foolishly, but, being a man of extremes, it seemed the right thing to do. Some people get seizures and go into convulsions when they stop drinking abruptly, so it's not a wise course of action without medical supervision, but I didn't.

Later, when the compulsion, the obsession, to drink began to leave my life, I sometimes wondered why I ever did it at all, considering that in sobriety and in full possession of my faculties, I was perfectly capable of being absolutely cuckoo on my own.

Later that year I would go off to Hollywood to make the fortune, but instead got a solid grounding in AA, as I got to a lot of meetings. 'Twas there I met a sponsor who made me laugh. A sponsor is another member of AA who, having been working at the task of staying sober longer than yourself, can give you the bit of advice when you need it. Jimmy had a face like the unsculpted side of Mount Rushmore and a voice with all the melodiousness of a cement mixer, but he had a powerful spiritual message. In the course of a conversation, it turned out that he had burglarized a saloon I'd been a partner in many years ago. Since then, he had spent more than a decade behind bars. Now, though, he was sober and a stalwart member of the fellowship, which goes to show what a strange grouping AA can produce, when your burglar becomes your sponsor.

The first year of sobriety was a difficult one for me, and I didn't know if I could keep going. But, over time, with prayer, patience, and persistence, a very decent, chummy sort of God popped into my life, and I've had chats with Her and remained sober for the past fifteen-plus years.

The same year we scattered Angela's ashes, our father died.

Throughout the eighties, word came of relatives of the previous generation dying in Ireland. My aunt Aggie and my uncle Pat, siblings of my mother, died, as did various siblings of my father. As they dropped off to eternity, wherever that might be, I was not very moved or grieved. There was an infected cavity of rage in my soul that still throbbed with pain. So, when news came of another relative being slid into the grave, there was hardly room in my own head for another thought on the matter.

I knew my father was living in Belfast, in housing provided by the government. He lived rent-free, with utilities provided and some food delivered, and apparently some neighborhood ladies thought he was a very nice elderly gentleman who enjoyed a chat. Over the years, there had been driblets of news about him—he'd calmed down, stopped drinking, was living the quiet life (which was not to wonder at, as he was well into his eighties, when the sap is very slow in the old tree). How this mad, eccentric lifelong alcoholic could have survived the battering he gave his body has to be a scientific wonder. He was almost yogic in his disdain for the comestible, and anything potable, such as tea, coffee, or water, could stay untouched for hours at his side. In his day, of course, he could put the disappearing act on whiskey and porter that would make the Indian rope trick seem ordinary.

Thinking about him one night, I worked myself into one of those self-pitying rages about what had been done to me as a child due to the father's desertion. I dwelt on the nights of hunger and days of desolation and watching for the telegram boy who might bring the money order from my father. I thought how, as darkness fell, we told each other that they were so busy that they were working overtime, and we went to bed, hoping again that in the morning . . . but they never came, and we were left shamed and humiliated and filled with fear by the disease of poverty.

One day, back then, when a gaggle of us ragamuffins were playing near the lane, one of the priests, in a rare move, stopped to speak to us. He was patronizing, as many of these jumped-up, bogtrotting yahoos could be, and asked us what we were going to be when we grew up. There was the usual chorus from a group of small boys: a fireman, a policeman, an engine driver, a jockey, and then he turned to me and I said I was going to America.

The man went cuckoo—he went the color of a polluted sunset, his neck came plopping like a frog's over his Roman collar, and between the clenched teeth he bit off a series of words condemning America. It was overrun with non-Catholics, atheists, Jews, Protestants, and anarchists, he said, and nobody went to mass, and children were not getting baptized, and 'twas no place for a good, God-fearing Catholic boy and, furthermore, if God had meant me to be in America, he would have put me there. Without trying to be the smart aleck, I said that God must have wanted me there, then, as I was born there. The priest stiffened and straightened himself as if he had just been gored in the arse by a long-horned bull, opened and closed his mouth a few times like a stranded halibut, and strode off.

Now here I was, a grown man in America, that sinful place

the priest had spoken about, and turns out he was 100 percent correct, as my life can confirm. And here I was, still moaning inside about the miserable childhood.

"It's got to stop somewhere," sez I to myself. I'd been attending the meetings having to do with alcohol, and it was beginning to dawn on me that I had been treading the same path as my father. What an astounding bit of knowledge it was to find out that there are more ways a man can desert his family than by just walking off into the night. I wasn't a great participant in the arenas of child-rearing, finding schools, attending teacher-parent conferences, or taking young ones to libraries, theatres, and museums. Like my own father, I wasn't a violent man at home; outside of a salutary whack on the arse of the young ones on occasion, I wasn't a beater of children or of Diana.

But, in certain ways, I simply wasn't there emotionally and spiritually. The analysis of how I got that way is relatively simple. When you are young and all the people you love and trust either let you down or die, and when the things you become attached to are taken away, you begin closing down the emotions, and you distrust any and all attachments. I didn't know that, so I just thought it was normal to be unmoved by death, disease, and tragedy.

Although I thought I loved Diana and my children, it was quite a long time after getting sober before I was able to tear down the solid Plexiglas and really let feelings for them penetrate my heart.

And one day, during my ruminations on the dismal state of my life, the world, show business, broadcasting, my relationship to my family and to Diana, etc., I decided to write a letter to my father. In it, I laid out all the travails of my life that were the result of his abandoning us, my feelings of inadequacy because of my lack of formal education, my not being able to concentrate on a job or task, the disrespect I have for time (other people's, that is), the nagging low self-esteem that plagued me, the need to drink in order to be comfortable in social situations. It was a very long and fairly thoughtful letter, in which I didn't really attack the man, but explained the consequences of his actions.

There was always a vagueness about the man, as if he didn't hear what you were saying, and he figured that he should be respected because he was our father, regardless of whether or not he acted like one. He placed great store on outward signs of piety, like rosary beads, prayer books, holy medals, and scapulars, and I remember one time seeing him sitting on the stairs, drunk, barely able to lift his head, saying the rosary and fingering the beads—he wouldn't see anything bizarre in that behavior.

As I wrote that letter, I grew more and more angry, not because of what he did, but because that most revolting of all human emotions, self-pity, began to take me over completely, causing the tears to flow, the sort of blubbering I sometimes succumbed to when I was drinking. I was told at the meetings I attended that there is hardly anything as powerful as self-pity to bring a relapse in the drinking department, so I purposely stamped it out and regained my equanimity.

To my astonishment, at the end of that letter, I was inspired to write that, for all that had happened and been done, I had no choice but to forgive him, because if I didn't, my own rage would do me terrible damage. I signed it, "Your son, Malachy." I couldn't bring myself to write "Your loving son."

I sent it off to an address in Belfast, and felt relieved and cleansed at having accomplished that bit of business.

A few weeks later, the letter came back, marked in pencil on the envelope: GONE AWAY.

I swore and cursed and raged again, that once more, even in his old age, he couldn't sit still and be there to receive my outpourings. With all due deliberation, I wrote another short note and put it with the letter in a bigger envelope, and bunged it off again, writing on the outside: PLEASE FORWARD!

A few weeks later, I got a reply from him. He had been in the hospital, and one of the neighbors, for some reason, was sending back all his mail. As he'd had a slight stroke, he was writing a very short letter, he said. As always, the handwriting was copperplate Victorian, and the content was quite formal. He wrote that he understood my anguish, and that he would endeavor to make it up to me, and thanked me for my forgiveness, and finished in one page, because he was rushing to catch the last post.

Here was a man that I'd not heard from in over twenty years, and he was rushing to get a letter mailed, a letter that could have waited for the first, second, or last post of the next day. Little did he or I know that indeed it would be the last post, for he died very soon after that.

Still, though, he had answered my letter, and it was clear to me that he had heard what I had to say to him, and cared about how I felt. It didn't make any of what had happened go away, but in a small way, it brought me a bit of peace.

At various times over the preceding years, my daughter, Siobhan, and my sons Conor and Cormac had been able to visit my father. Though there was a bit of difficulty in communicating because of his Belfast accent and somewhat impaired memory, I was glad that they did get to meet a grandfather who was pleasant and affectionate, so they didn't have to rely on my bitter memories.

When word came, there was a gathering of the McCourt clan in Belfast for the funeral, which I didn't attend. My brothers Frank and Alphie went and did the honors as the oldest and the youngest members of our crowd. Frank tells the story of going into the death house, the little mortuary close to the Royal Victoria Hospital, and seeing our father laid out in the coffin with a bow tie stuck on his collar; his false teeth hadn't been inserted, so he looked like a seagull. So as Frank knelt on the prie-dieu beside the coffin, he began to shake with laughter at this absurd little man in the coffin, and the relatives mistakenly thought Frank was overcome with grief.

In keeping with my father's penchant for surface propriety and external piety, he left several thousands of dollars for a fairly elaborate tombstone to be erected in his memory. Perhaps he thought he wouldn't be remembered for his wasted, cursed life. Probably, he didn't know that tombstones were originally boulders rolled onto graves, intended to keep the dead from coming back in anger, since they now knew the secret lives and thoughts of their apparent loved ones.

Though I tried in the spirit of forgiveness to grieve for that man who fathered me, not one tear squeezed past the eyelashes. The best I could do was to wish him well, wherever his spirit landed, and to hope that he wasn't in that hot place, where he could light his pipe with the tip of his finger.

Forgiveness or not, Frank, Malachy, Michael, and Alphonsus McCourt had now joined the ranks of grown-up orphans, and were now on the front line of life. We're next. The prospects of behaving forevermore in a mature, adult way didn't appeal to me at all, for wasn't I the social gadabout with the laugh on the lips, the quick quip, and the huge capacity for whatever alcoholic beverage was on hand? But death doesn't take note of what you were; it simply puts you on the list as a client, and you have to utilize the services provided, whether you like it or not, and whether you want them or don't.

I envied folks who went to parental funerals and told stories of their beloved dad or mom, of their kindnesses, their peccadilloes, of their love, and of their eccentricities. I wished that I could have done that, but it would have been a false and hollow eulogy.

There is the story of the death in an Irish town of a dreadful man. He was wealthy, but a miser who lived alone. He lent out his money at usurious rates, and would seize collateral property if the loan were a minute late. A more despised man could not be found, but he left all his money to the church. There was an elaborate funeral with High Mass, and a hypocritical, fulsome eulogy from the parish priest, who was the beneficiary of the will.

Many of the neighbors came to celebrate the miser's death, and when the priest asked if anyone else would like to say a few words about the deceased, there was an embarrassed rustling, but no volunteers. Finally, the priest pointed to one senior townsman and practically commanded him to speak. The townsman shambled up to the lectern, cleared his throat, looked around, and said, "His brother was worse," and returned to his seat.

In my father's case, I could perhaps have stood up and said the son was worse.

Part III

How It Works

If there's one thing the Brits do well, it is centralizing all facets of show business. In London, they have the West End for theatre, the BBC and others for television, and the big studios are all within driving distance of the city. But here, the actor is driven gaga by having theatre and some television produced on the East Coast, and all the movies and filmed television done on Da Coast. You can't make a living in theatre, so you dash to Da Coast, only to sit around there for months awaiting the call.

Some friends of mine offered me a substantial loan without interest, and though I practically pledged an oath to Diana that I would not borrow any more money, run up any more bills than we had, or attempt to get new credit cards, I took the loan. Even with my newfound spiritual path, I couldn't forbear telling the lie and behaving in the same old way. I thought I could conceal the money and just barrel off to California, so I made up some story about having to go to Hollywood on a very big audition. My plan, once and for all, was to get a series, make a vast fortune, and retire to a life of leisure.

About seventy thousand actors in Hollywood had the same idea, but, of course, I'm different.

John and Marcy Carsey provided me with very beautiful accommodations in their house in Malibu. I secured a second-hand car held together by bungee cords, and rang up a very able woman named Joan Scott, who owned a successful agency called Writers and Artists. She was kind enough to sign me up, though I think the other folks in the agency thought she had taken leave of her senses.

One promise you have to make when signing on with a West Coast agent is that you intend to move there with ye olde lock, stock, etc., which I did, of course. Some old New York friends threw me a Welcome to L.A. party—Jack Riley, Pat McCormick, his sister Mona, and Chuck McCann—and I was now an accepted member of the circle of New York expatriates.

The big push was on. But as John Carsey said, "You go out there, sit by the pool at 10:00 A.M., doze off, and you wake up at noon and you're ninety."

Work for me was scarce, as there wasn't a great demand for chunky actors in their mid-fifties with white hair and an Irish accent. My Hollywood venture, and my plan of getting the big series and all the money that goes with it, soon fizzled out, along with the money. After I'd stretched the hospitality of the Carseys to the limits, I moved out of the Malibu house into a very dark and dingy apartment, which did little for my spirits.

Writers and Artists agency would seem to have forgotten my existence, as I very rarely heard from them; more often, I had to call, only to get a "There's nothing doing" message from a bored agent who needed to get off the phone to attend to more glamorous clients than myself.

The old joke that changing agents in Hollywood is like changing deck chairs on the *Titanic* is as accurate as it was the day it was coined. There is no point in actors' calling agents or managers for news, because those folk work on commission, and if there is a dollar to be made with an actor, they will be on the phone to the hungering thespian tout de suite.

There is another ancient saw about agents, which was coined by Fred Allen, to wit: You could take all the sincerity in Hollywood and fit it in a flea's belly button and still have room left over for six caraway seeds and an agent's heart.

They come in for quite a thumping, agents, and many times undeserved. Agents have to be accountants, baby-sitters, psychologists, and therapists, and have to listen to many moaning idiots who have no business being in the acting business, especially when their one talent is delivering pizza. (Hollywood being what it is, though, a place where talent counts for little, the pizza deliverers sometimes end up getting the part.) Most agents work very hard to get work for the actor, and they get paid the 10 percent only when they succeed, plus they have to comfort the vast array of untalented folk they are forced to represent when, as usually happens, they are turned down for a job. The conversation usually assumes this form:

"Hello."

"Hello, Tad, this is Mitzi at the agency."

"Hi, Mitzi. I guess you're calling about the part on *The Milk Never Sours*. Boy, that was some audition I did. Really blew them away, I did. So, when do I start?"

"Well, Tad, they really loved you and would have used you, but they are going in another direction with this part."

"Oh!"

"But don't worry. You'll get the next one. . . ." And so on.

With all that failure surrounding me, I hardly needed a beacon flashing on my chest at auditions to proclaim "I am a loser!" I was surprised at times to find myself competing with very well known character actors, some of whom had Oscars, Tonys, and Emmys, and every other award you could think of. It wasn't that they needed the money, but they needed to be working to be acknowledged, and in the so-called twilight of illustrious careers, they were once again being treated as if they had never set foot on a stage or set before. But still, they were willing to endure the rejection that is 95 percent of an actor's life.

It has taken me until fairly recently, in my sixties, to formulate an emotionally healthy, spiritual way of going about the acting business. The first item on a spiritual agenda is that I am not what I do, that I am what I am, and therefore I am a walking miracle, on a daily basis.

Another bit of wisdom I came upon was that part of my daily work was looking for a job, which, though nonpaying, is still work, so I'm always working.

Then there is this: Acting is not a creative art in the accepted sense; it is a carry-the-message kind of thing. More often than not, the messenger cannot be any old delivery boy or girl, because for the message to get through, it has to be delivered in some sort of code, and not everybody possesses the code for a particular message.

Unfortunately, the people who set themselves up as selectors of the various messengers are generally incapable of delivering or carrying any messages themselves. Indeed, from an actor's perspective, many of these casting folks have all the physical deficiencies of baseball umpires: They are apparently sightless, hearing-impaired, plus lacking in intelligence and sensitivity.

I came to believe that if a divinity were to allow these people a glimpse of my bright talents, they would be totally blinded for all time, so in His mercy, He allows them to remain in ignorance.

It's not that they aren't nice, the casting people. They are generally friendly, and they never seem impatient or hurried. Niceness is the most fearsome weapon they have, for there is no defense against it. Back then, there were times when I wished they would just say, "Look, McCourt, you are all wrong for this part. As a matter of fact, you are all wrong for everything, so why don't you fuck off out of here and don't be wasting valuable time, which could be used to interview people with talent."

The only time I came close to something like that was the audition for the Irish priest, when I was told my Irish accent was not an appropriate accent for someone from Ireland. There was also another time, on the set of *The Brinks Job*, when the director Billy Friedkin said to me after a scene, "That was about as bad a piece of acting as I've ever seen. Would you mind putting your intelligence to work here, so we don't have to put up with this kind of shit." This is about as direct as you can get, but it is Friedkin's blunt style.

O ne day, whilst I was collapsed in a depressed state in my dingy, dark apartment, the phone rang. I let it go on ringing and ringing until the answering machine did its job. A very Germanic voice, which identified itself as something like "Schultz Kiel," asked if I would call him back, as he was in town on a project and my name had been given to him as someone who could help.

I forced myself to pick up the telephone and acknowledge that I was there. Mr. Kiel said he was a producer collaborating with John Huston on a film version of the James Joyce short story "The Dead," and would I be interested in being a part of it? Damn right I would, as it's a wonderful story, though there is not a great deal of action in it.

The spirits soared at the prospect of work, and not alone work, but participation in a beautiful classic that would now become a classic film. Kiel had asked that I meet him at his hotel, and with the heart very warm and cheery, I set out in the broken and battered Toyota for what I felt was going to be the turning point in my career. The reception was most cordial, and I was informed that there was going to be a cast of great actors: Siobhan McKenna, Donal McCann, Dan O'Herlihy, Donal Donnelly, and more. The genial German said that there was a very good part that I would be ideal for, and that he would ring me about an appointment to meet with the great man himself, John Huston.

Then he asked me if there were any other Irish actors in Hollywood he should know about, as Huston was insistent on casting only Irish actors. I gave him the names of several: Helena Carroll, a gifted woman of the theatre, and cofounder of the Irish Players in New York, where I'd gotten my first acting job Off Broadway; Tom McGreevey, a lad I'd met when we were both working on *Ryan's Hope*; and Maria Hayden, a young and beautiful girl who had recently come to California from Dublin and was experiencing the usual struggles.

It was a warm and jolly meeting, totally unlike the usual audition, as Mr. Schultz Kiel was an erudite and well-read man, with a European contempt for the shallowness of life in this Tinseltown and its anti-intellectualism. We agreed on everything—the greatness of Joyce, the genius of John Huston, the talents of the marvelous cast, and how grand the whole film project was going to be. We shook hands, and he said he would ring the next day, Friday, about a meeting with the director, John Huston.

I got into the old clunker car, which didn't look too bad now, and set off for the home base, trying the air of "The Lass of Aughrim," and then singing any old song that was lively. I thought of having a celebratory drink, and decided that would be a mistake, considering what other celebratory drinks—and drown-my-sorrows drinks, for that matter—had done in the past. Sobriety can be a pain in the arse at times, particularly when a drink seems to be in order.

Instead, though, I did as I'd been told I should when such impulses arise, and went to a meeting. I listened carefully to more experiences of the devastation caused by drink, and other tales of triumph over addiction and obsession. Through it all, the high spirits of prospective work stayed with me, and when I went home, it struck me that my apartment was not so dingy as I had thought.

Next morning, I awoke with the birds and got up and did the stuff I'd been neglecting, like gathering laundry, washing the dirty dishes that had piled up, tearing the sheets off the bed and replacing them with crisp clean ones, sweeping the floor, and throwing out the piles of periodicals—*The Hollywood Reporter*, daily *Variety*, and *L.A. Times*—that were reaching printed heights never before scaled by humankind. I decided that a swift foray to the Laundromat would not torpedo my chances of stardom, though it did cause me anxiety to be away from the phone.

When I did return an hour and a half later, the red light on my answering device was still and unblinking; there had been no calls in my absence. Relief and disappointment took turns in the brain. I popped in and popped out, always looking for the blinking red light, but it failed to blink, and continued staring at me like the eye of a baleful weasel. I thought of initiating the call myself, but fear of being thought overanxious stopped me.

The night came, and I was faced with the prospect of an interminable weekend, with a vague promise and no commitment. Under my normal modus operandi, it would have been the dash to the saloon, but that was no longer an option. So, instead I related the problem to some of the friends, anyone who would listen to a whimpering actor who was waiting for THE CALL.

The weekend passed as all weekends do. Monday dawned, and the little red eye remained unblinking. Reducing the pride to minus-one, I finally telephoned Mr. Schultz Kiel, who was not in, so I left a message that I called. Tuesday arrived, with no call again, so I telephoned a bit irritably and left another message, to which there was again no response. Now I was getting angry, and on Wednesday called three times, and finally nailed the honorable, upstanding man who despised the chicanery and dirty dealings of commercial Hollywood. He asked, a bit impatiently, what I wanted. I said, "What about the appointment with John Huston?" He said that the part he had mentioned to me was already cast, and that he was extremely busy, good-bye, and he hung up.

A demolished Malachy McCourt sat staring at his telephone, not wanting to credit for one moment what he'd heard. If you want to give God a good laugh, tell Him your plans. Did I hear Him chortling in the heavens above?

I felt as if I'd been rammed in the solar plexus with a flagpole borne by a company of Marines. I crawled into my bed and covered my head to shut out the world and lay there whimpering like a beaten baby.

I'd been rejected at other times in my life, and to this day I don't know why the loss of this one job reduced me to such a mess. In my memory, there is only one other time, outside of the departure of my first wife with the children, when the pain was that palpable.

When I was about six, my grandmother announced that there was to be an all-day excursion to Galway on the train, and that I was one of the children chosen to go. She said it was by the sea, and there was a strand, and sometimes you could see ships sailing past. I'd get to eat a sandwich on the train, and if I was very good, I could have a lemonade made in Limerick by Olo, a very holy company, because Olo meant "Our Lady's

Own," in honor of Jesus' mother. There was talk of chocolate and Cleeve's toffee, and maybe cream buns, and when we got to the seaside, there were men who sold ice cream between wafers or in the cones.

The days dragged on for a six-year-old, as the trip was two months hence, and there wasn't an adult within speaking range who wasn't bombarded with questions about the train. How fast would it go? Would you see fields with cows in them? Could a bull knock a train down? Could I see the engine and the man shoveling the coal into the fire to make the train go faster and faster? Would the driver pull the cord to make the whistle blow very loud, so that everyone would know we were coming and get out of the way?

Every night for two months I dreamt of the sea and jumping into it, because they thought that you can't drown in salty water, 'cos you float. Someone showed me a picture of a man lying on his back in the Dead Sea, holding an umbrella to shade him from the sun, and he was reading a book and not sinking. When we got to Galway, I was going to get a new pair of sandshoes, gleaming white, that let you run as fast as the wind, and it wouldn't matter if you got them wet in the sea, as they dry so fast.

No journey to the seaside was complete without a bucket and shovel for the building of sand castles, and I was going to build the biggest one in the world, with towers and windows and a moat to keep the enemy, whoever they might be, from crossing over and invading, and not even the waves would be able to knock it down, it would be so strong.

The grandmother was not a nice woman at all. I never saw her laugh or smile or be funny like the other boys' grandmothers, and she was as likely to give me a clitter in the ear as look at me, so it was wise to stay out of reach of her swift right hand. When I was six, she was about sixty-five, but she seemed so bent and ancient and stooped that I thought she must be at least a hundred. I didn't care now, though, because, for the first time in my life, she had done something nice for me by saying I could go to the seaside with her and the other children. The other children were my cousins, the Sheehans, whom she liked better than the McCourts, 'cos their father was her son. My mother said that my grandmother thought the sun shone out through her son's arse.

The morning of the excursion, I got out of bed around four o'clock. Off in the dark I could hear trains shunting back and forth in the yards and, from nearer by, a peep or two from the birds. The dawn crept up in a blue-gray haze, and I went over to my mother and father's bed and shook them. "It's time for the train," sez I.

"In the name of Jaysus, will ye go back to yer bed and not be waking us up in the middle of the night!"

"The train is leaving," sez I, "and it'll be gone by the time we get there." My mother was always late, and if you rushed her, she'd say, "When God made time, he made plenty of it." But this day when I'd see the green sea that changes its colors to blue, and has ships sailing on top of it, and stretches so far you can't see land, was too important to be late for.

She got up and lit a fire with some sticks and boiled the kettle and made tea and had a bit of bread and butter. Frank got up and got dressed grumpily, because he was always the first to get up, but I beat him this morning.

We left our lane and walked down to the lane where my grandmother lived, where the cousins were all dressed and jumping up and down with excitement. I could hardly talk with the way my chest felt, and I wanted to sing and shout and laugh out, and I wanted to cheer and say silly things, like "flupidle, addeloodle, diddley pink sam, aram, bonkbinkle, ooramaday," and I loved everybody, and I loved the world, and I even loved my grandmother.

This band of people, consisting of my grandmother, my mother, myself, Frank, and three or four cousins, moved out of the lane and up the hill toward Barrington Street, and there standing at the top of the hill was my cousin, the eldest of the Sheehan clan. He was staring into the distance as if he didn't know we were coming, and then my grandmother noticed him.

"What are you doing there?" sez she.

"Nothing," sez he.

"Come with us," sez she.

"Awright," sez he.

Then the grandmother turned and, pointing a finger at me, said, "Go home!"

I looked around, thinking she had seen a dog behind me, but there was no dog and she was talking to me. I looked to my mother for some sign that she wouldn't let this awful thing happen to me, but she looked away and shrugged helplessly. They walked off toward the railway station and left me standing alone at the top of that hill with a lump so big in my throat that I couldn't even sob. The tears cascaded down the cheeks, blinding me on that sunny morning in June, as I made my way back to our house in the lane, crying and crying and crying all day, thinking of the sea, and the sand, and the sky over it all. I would never have a chance to see it, I thought, in my whole life. I couldn't be comforted, no matter what my father did to cheer me up. I didn't get to see the sea 'til I was about fifteen and on my way to work in England, and it wasn't the same at all.

Over sixty years later, and here I am, a grown man, whimpering about it again! Jesus, will I ever grow up?

I stayed in my bed in my dingy apartment for a bit, but they had told me at my meetings that this, too, shall pass. I got up and went about my business again, listlessly moving from audition to audition, with no result except a bit part on a soon-to-be-canceled television series.

The phone rang one morning, and I picked it up to hear the familiar Teutonic voice of Schultz Kiel telling me that there was another, smaller part in *The Dead*, and would I be interested? The need to work and get income, and my raging, wounded, pride at the previous rejection, got into a frightful battle. It was a long moment before I could conquer the pride and say, "Yes, I'd be interested."

"Goot," he said. "Please come on Sunday and ve vill meet John Huston and make ziss matter final." And the world was smiling again, as another actor got a job and wouldn't be in the way of the others still seeking work.

Sunday came—an odd day for meeting a director, but *The Dead* was no ordinary production—and I hied my way to Schultz Kiel's hotel and got into his car with several other actors, and up into the Hollywood Hills with us. We were shown into a study-cum-library type of room, with low lighting, containing a long table. It also contained John Huston, who sat at the head of the table and waved us to be seated. He was an impressive man in the films, and even more so in person, with his distinctive voice. A genial, friendly man he was. His breathing was not without effort, and he was using oxygen to make it easier. I assumed he was suffering the emphysema from the heavy smoking in which all glamorous Hollywood people indulged.

He didn't ask anyone to read anything, and, as we were all Irish, the conversation meandered about Ireland, its economy, its contribution to art, and how the film industry there was growing. He reminisced about his day at the Galway Hunt, and the tough terrain one had to ride in pursuit of the cunning Reynard. Finally, this gracious man thanked us all for coming, wished us a Merry Christmas, and told us he was looking forward to seeing us at the start of filming on January 3.

Handshakes all 'round, and off we went to our respective abodes. As everybody in the cast was receiving minimum Screen Actors Guild salary, about two thousand dollars a week, and it was a guaranteed twelve-week shoot, I figured I'd have twenty-four thousand dollars to start the year, which was quite a boost. Some of my debts would be paid, and all the usual planning flowed through my head in anticipation of this bit of good fortune.

When I arrived back at the apartment, the little red eye on the machine was blinking benevolently, indicating that several messages were awaiting my attention, and attend to them I did. A couple of friendly calls, hoping that all went well with the Huston interview, and one from the redoubtable Mr. Schultz Kiel. Would I please ring him back, as he had a question for me.

I rang him back, expecting to discuss some details about my part, or the shooting schedule. He answered the phone on the first ring and, without any preliminaries or formalities, said that John Huston had made a mistake. There was no part for me in the film.

For a swift second, I thought the man might be indulging in some heavy Teutonic humor, but 'twas not so. He was serious.

A murderous calm enveloped my whole being as I took the requisite breath and asked how this could have happened. The man didn't know, and informed me he was busy and had to leave for an urgent appointment, and then the phone went silent.

I didn't believe what I'd just heard. It seemed to me that every other Irish actor in Hollywood, including those I'd recommended to Herr Schultz Kiel, was working on this epic. This couldn't be happening.

I decided the best thing to do under the circumstances was to get langer's drunk. The dingy apartment now had all the cheeriness of a used coffin, and I thought briefly of systematically destroying the place as well, but as I had no ax, I couldn't do a good job of destruction. Instead, I left to find a bar. All this sobriety business, all this being moral, and truthful, and faithful, was not working. If there was a God—and in light of my experiences, there wasn't—he sure as hell was playing some bizarre game with my life and expectations and, as the man said, I was mad as hell and I wasn't going to take it anymore.

I'd noticed a run-down sort of place on a side street, and I headed straight for it. It was not quite as run-down inside; it was quite respectable, in fact. I parked myself at the bar, trying to remember how to act at bars, as it had been over a year since I had sat at one. The barman was busy when I got there, so I had time to survey the labels on the array of bottles. To my utter and complete astonishment, there, in pride of place, was a bottle of one of my favorite Irish whiskeys, called Black Bush, a beverage made by Bushmills of Northern Ireland. This is perfectly providential, sez I to myself, as I won't have to drink rotgut.

A genial California barman came to my side, asked how I was, and what could he do for me. One of my standard responses to "How are you?" is "If I was any better, God would be jealous," which amuses people and makes them think I am a very clever fellow indeed, so I responded to the lad's query with that little witticism. I averred as how I was delighted to see evidence of the management's elegant tastes in carrying Black Bush in the shelf stock.

He looked a bit puzzled 'til I pointed the bottle out to him. "Never served that before," sez he. I told him now was as good a time as any to begin, and he should heft out a double with no ice, but I'd venture a glass of water on the side. The lad complied, and I put a ten-dollar bill on the bar.

I sat and sipped at the water, raging within at the cruelty of this stupid, awful, fucking acting business. I at first convinced myself that John Huston had set out to torment and torture me, and then settled on the saner thought that the man doesn't know me, it's just that his emphysema and other illnesses were impairing his judgment.

Nevertheless, the whole fucking thing was driving me to drink. I hadn't had an alcoholic drink in a year and a half, since July 22, 1985, to be precise, and here I was letting this man, who had only met me once, cause me to pick up a drink.

I recall scoffing at people who defended their crimes by saying, "The Devil made me do it." Was I going to say that John Huston made me do it? I suggested to myself 'twould be a good idea to go to the lavatory and look at myself in the mirror, which I did. I was standing there chatting to myself, when another man came in, which put a stop to the chatter. I further suggested to myself that I go outside and take a walk.

The barman was looking a bit quizzically at my untouched whiskey, the slightly sipped water, and the untouched ten-dollar bill on the bar. I mumbled something about having a blah blah blah in the car, and I would be back presently. I went and sat in the car for a bit, walked around for a bit, got back in the car, and decided to drive for a few blocks to distance myself from the amber beauty of the Black Bush.

Then a whole convention of rowdy, mad demons started their proceedings in the confines of my skull; not alone was every state represented, but every country too; and there were loud bands, rock and brass, punctuating the speeches with blaring music and drumrolls, a cacophony of lunacy known only to the alcoholic. There were motions proposed and seconded and passed without demur that I should go back to that bar and claim the drink that was rightfully mine. Plus, I'd left a ten-dollar bill there, and there ought to be change from that. And furthermore, did you not deserve a drink after what had been done to you?

So, fuck the program and their God and Schultz Kiel and John Huston and Hollywood and the film business and the acting business, not to mention any business, and don't forget, fuck the world, too, while you are at it, and all the arseholes who inhabit it.

I stopped the old, broken-down Toyota and parked, and reached for a cigarette so I might at least have the comfort of my other addiction, but there wasn't one in my vicinity. Christ! I must have left the pack on the bar, along with the drink and the money. "Another reason for going back!" roared the demon conventioneers in my noggin.

I returned to my apartment and decided to make a desperate call for help, and, as I flipped through the pages of the address book, I encountered the name and number of Fionnula Flanagan, a lovely and compassionate woman, well known for her talent in the acting trade. Along with her husband, Dr. Garrett O'Connor, she was known to be a wonderful adviser for actors heading off the deep end. This would serve a fine description of me at that moment, and, fortunately, she was on the premises when I rang.

In retrospect, 'twas much ado about fuck-all, but the lady listened as if I had suffered a great loss in my life. In passing, she mentioned that John Huston had called her to say that if his daughter, Angelica Huston, hadn't asked for the starring role in *The Dead*, it would have gone to Fionnula, as she had done the definitive interpretation of all of Joyce's Women in that magnificent film.

I acknowledged that she must have been devastated, and then continued right along with my own barrage, as if I were the only wronged person in Christendom, Hebrewdom, Hindudom, and Muslimdom. Fionnula listened and listened, and when I'd done with my whinging and whining, she buoyed me with a gentle and spiritual talk. She assured me that, no matter what some people do to you, you are still loved by God and by those who know you, and all you have to do is let it into your heart. The business of acting will murder you if you allow it to, if you look at it the wrong way, she told me, so it's essential to love yourself and start from there, and you won't go wrong, no matter what is done or omitted.

Her soothing voice quieted all the roaring demons in my head, and the compulsion to return to that bar and reclaim my drink faded away. Instead, I betook myself to a meeting and congratulated myself on my very narrow escape from the bottle.

Christmas was approaching, and the songs and carols of the season were blaring from radios and loudspeakers in the shops. As it was with many another person, this was the time of year that most drove me to gloom and depression, and the music didn't help much: "I Saw Mommy Kissing Santa Claus"; packs of dogs woofing "Jingle Bells"; nonsense about winter wonderlands; not to mention "Christmas in Killarney" and that dreary song, "White Christmas"; it all combined to slide me into a slough of despair, wishing it would all go away.

It was nonetheless necessary to get myself home to spend the holiday with Diana and the family. It wasn't a jubilant time, as John, my father-in-law, was ailing and had seemed to age very rapidly. It was a tense time also because Diana was not at all happy with my absence in Hollywood. She didn't understand what the hell I was doing out there, and I couldn't give her a good explanation. Telling her that I'd been deprived of one part through odd circumstances, and now I'd surely be rewarded, didn't even sound reasonable to me.

Despite the tension, or perhaps because of it, I returned to California and resumed the tedious auditioning, driving from studio to studio, from casting office to casting office, all to no avail. "What in hell am I doing here, making a fool of myself in front of children posing as adults who have the power to give me a job for a week or two?" was the question I most often asked myself, and yet the insanity continued.

My son Malachy, who had heeded my advice and betook himself out of the environs of the East Side of Manhattan, where trouble reigned for him night and day, was living and working in San Francisco. My brother Michael had helped him to settle in and get a job in the saloon business. When he came on a visit to Los Angeles, he was aghast at my living conditions, but I pooh-poohed his concern by ensuring him it was a temporary accommodation.

One Sunday I received a phone call from a woman who said she was a casting person looking for people with genuine Irish accents to do some dubbing for a nearly completed film. Dubbing is the taping of another voice when the quality of the original voice recorded for a movie is muffled or garbled. I said, joking, that it must be John Huston's *The Dead* she was talking about. To my great bewilderment, she said that it was. She told me that my name had been given to her, and she wanted to know if I knew of any other Irish actors. I mentioned a few, and gave her their phone numbers, and then she asked if I was interested in doing some dubbing?

Again, the angry pride rose up inside me and told me to tell her, "Fuck off!" while the prudent, sober side said, "Cool it. You need the job." Once more, the prudent, sober side won out. I choked down the angry words and told her I'd do it. She said she would ring me in the morning, and was delighted to get me, as John Huston was such a stickler for having the genuine Irish voice that he wouldn't ever hear or audition people who were not the real thing.

"Ironic," sez I to myself, "that I was going to be some of the 'real' Irish voices in this movie."

The dawn arrived as usual. I got up and did the morning

ablutions, and tossed off a muted prayer, making sure that no one heard me, and then proceeded to wait for the lady of genuine Irish voices to ring, with directions to the studio. The hours passed, and that afternoon I picked up the telephone with all due deliberation, took several deep breaths, and tapped out her number.

When I got her on the phone, she cheerfully asked what she could do for me. I said she could tell me where and at what time the dubbing session was to begin. She said that they had completed that work this past morning.

"Didn't you ask me to do that?"

"Well, there was no real commitment to you. I was just ascertaining your availability, and we ended up getting a wonderful actor to do the voices," and then she mentioned his name.

"But he's English!" I bellowed at her.

"Yes," she said, "but he does such a wonderful Irish brogue."

Speechless is not the word to describe my condition at that moment, but it's close enough. She was chattering on about the wonderful dubbing session, when I interrupted her with, "Will you be seeing Huston and Schultz Kiel again?"

"Yes," sez she.

"Tell them if they need any more bodies for *The Dead*, try the goddamn morgue, and for Christ's sake, leave me alone!" and I slammed down the phone.

A very large and impenetrable wall seemed to have been built in Los Angeles to keep me from achieving financial security, so I prepared to return to New York. However, another whack in the solar plexus awaited me.

I had, I thought, successfully hidden the source of my funds from Diana, but of course it came to light that I had borrowed again despite my pledge not to. Naturally, she felt she was deceived and, in a sense, betrayed as well, because the money came from friends of ours.

When I called and said I was abandoning California and coming home, there was no enthusiasm from her. She told me that the trust we had built up was now in ruins, and she did not feel we could live together as husband and wife without trust and the love that goes with it.

The cold, steel hand of fear and disappointment took a firm hold on the heart and began the powerful squeeze. Welters of emotion charged through me, as changeable as Irish weather—rage followed by sorrow, then anger, then grief, then repentance, then resentment. But I went on with the plans to return to New York. My confidence was gone, the self-esteem had totally disappeared from my charts, and faith in any kind of reward for redemptive efforts had now become a matter for scoffing, but I still resisted the call of the whiskey.

Back in New York, my son Conor was, at that time, sharing an apartment with a grumpy fellow who did not take kindly to the idea of a roommate's father occupying any space. My sojourn there was brief. Conor was most apologetic, but the other fellow was the lessee, so there was nothing he could do.

A few phone calls, and I got in touch with Fran Holland, a friend who shared an apartment on the fringes of Harlem. It was a fourth-floor walk-up, and by the time I'd lugged my bags up the stairs, I was ready to settle down, not move again, but the social contacts had to be made, so that people would have the dubious privilege of inviting me to lunch or dinner, or to whatever was going on in the way of free food or entertainment.

My stay with Fran Holland was fairly brief, as the accommodations were limited, and my body in the living room was an impediment to anyone else living there. I contacted Adrian Flannelly, who was, and is, the host of a radio show with the largest Irish listenership in the United States, and he's been at it longer than just about anyone else on radio. A remarkable and oddly unsung man is Flannelly.

Two of his children, Eileen and Paul, were in an apartment in Queens that was quite large. With their consent, I moved into a small anteroom, cozy and private, off the living room. The next order of business was to scope out the meetings that keep one sober, and there was a plenitude of them in that area, since there was no shortage of Irish. Not that the Irish are more intemperate than anyone else in the liquor department. At one time, people laughed long and hard if anyone mentioned a Jewish alcoholic, but that has changed. The disease is as indiscriminate as cancer, and it will hit men, women, or children, be they sybarites, ascetics, or anyone else.

The *Adrian Flannelly Show* is presented every Saturday morning from 11:00 A.M. to 1:00 P.M., and on Sunday nights from 8:00 to 10:00 P.M. It's a mixture of music—Irish folk as well as traditional instrumental—and talk, with interviews of presidents and prime and other ministers. It is required listening for the Irish community.

I was invited by mine host Adrian to sit in and participate in many of the discussions of events in Ireland and stuff concerning the Irish in America. Heretofore, on my own radio show, I had studiously avoided things Irish, as I wanted my shows to be of a general nature, concerned with life in the U.S.A. Moreover, what I perceived as a right-wing bent in the Irish who had become successful was galling to me. In the past, I'd never lost an opportunity to goad and prod them on their amnesia about their humble roots. I must have been successful because, at times, the station would be hit by what appeared to be an organized letter-writing campaign, suggesting that I was a foul, rude, vulgar loud-mouth who had no place on any radio station. Boy, did I love that! These letters only provoked me into further denunciations of the self-righteous bigots who, until recently, enjoyed sexual activity primarily with their woolly lovers, the sheep of the fields.

But now, I was in a different position. Flannelly had attained prominence and permanence by dint of perseverance and for-bearance. He didn't much like many of the people he had on the show, but he is a fair man who believes all sides deserve a hearing. There were occasions when he was absent and asked me to host the show for him. If it weren't for the restraints imposed on me by his producer, Patricia O'Callaghan, a fine young woman of propriety, it's likely that the man would have returned to find his program had been canceled.

Of course, even biting my tongue to prevent the more out-rageous of my comments from escaping out onto the air, the complaints poured in. We dealt with them by implying that there was nothing that could be done to keep me off the air due to some obscure regulation. Some people offered to write to the government to remove me, but Adrian assured them it would do no good. Some people said we were descending to the level of Howard Stern and other such vulgarians, but we prided ourselves on the skewering of stuffed shirts and deflating pompous arseholes.

Though I was receiving rent-free shelter, I had exhausted the unemployment benefits, and there wasn't much money in the till to pay anybody. I looked in the papers, and called about jobs, but the only response I got was from an ad in the *Daily News*, looking for people to show apartments. I was given an appointment for an interview, where a girl (yes, she was a slip of a girl) looked at me as if I were a not particularly interesting antique. She said that showing these apartments entailed climbing a lot of stairs, as the buildings did not have elevators, and she offered the opinion that it would be too much for a man of my age and condition.

I didn't argue with her, and left deflated, certain that I was finished as a useful and productive member of society. That was that. I decided to apply for welfare. I analyzed the word: "to fare well, to journey in health." "Welfare" is a soft word, like a fluffy pillow, light and comforting. The reality is neither light nor soft nor comforting.

The Department of Social Services. The office with the evergreen walls and the fluorescent lighting, rows and rows and rows of women, mostly black, prisoners of poverty and drugs and alcohol. The workers listen patiently to desperate people with terrible, crushing needs, people who need a bed to sleep in, or a specialist for a very sick child, who need detoxification for drugs or drink, or just need enough hope to live another day.

When it's my turn, I try to make myself very small and unobtrusive. Godammit, why didn't I wear a cap and get some cheap sunglasses? But it's too late now.

Mrs. Bates mispronounces my name and misspells "McCourt" —she makes it "McCord." I correct her on both counts. She looks irritated, so I assure her that it isn't an easy name, nor too common a one. There was no apparent interest in discussing my name, as those triplicate forms were hard to correct.

There was a host of questions I had to answer about income and assets, about work history, and whether I was ever going to work again. What was my household income, and the income of my wife and children? What was my rent and my child support? Did I have a bank account? How much did I owe, how much was owed me, and what was I planning to do to collect same.

It was a wearisome task. I'd arrived there at 9:00 A.M. and it was by then 4:00 P.M., a full workday for most people. Of course, my first thought had been, "What if someone sees me there?" The same thought I had when I first went to twelve-step meetings.

When I was a child, we were told, "Do not take food from neighbors. They might think you need it," so we said, "No, thank you very much," when the mouth was watering and the stomach was making an effort to come out of my body and embrace the offerings. What if I needed it? What if they, whoever they are, saw me at a meeting? They might think I need to get sobriety, which they should have been thinking when they saw me roaring at some bar.

Yes, maybe someone would see me at the welfare office, and I'd be humiliated because I wasn't noble enough to be humble. I remembered my mother, Angela, standing with the other poor women, trying to maintain a bit of dignity in the face of contempt, waiting for a miserable few shillings to buy tea and bread and a drop of milk. The man shouted at her about concealed money from my father, 'cos you can't be up to these cunning people from the lanes of Limerick. You'd think they didn't have two ha'pennies to rub together, but they have enough to buy you and sell you. Salvo after salvo of insult to see if the woman would crumble under them. But, no. Her children were hungry and neither bird nor beast will allow the young to want, no matter the danger or the threat.

Nobody was threatening me or taunting me except myself, and my false pride was getting in the way of my need.

The woman at the Department of Social Services told me I'd have to go to Manhattan for a physical, and to receive a photo identification, the next day. She told me to be there at 8:30 A.M., and don't be late, or I'd have to come back another day and my benefits would be delayed.

I got there at the appointed hour and asked the security guard, a uniformed woman, about getting my picture taken. She gave me a very searching look and pointed to a wall, against which a group of people had lined up, and told me to get over there. She kept eyeing me as I tried to read my paper.

I was most uncomfortable under this hard scrutiny, until she came toward me, pointing a finger and saying:

"Now I got it! You're Kevin from *Ryan's Hope* on ABC."

Right! All eyes were on me. I had been spotted. I owned up to being the character on the soap opera, and she shouted to the workers behind the counter:

"It's him! I knew it all the time! We got a movie star here!"

The ground was not kind enough to swallow me at that

moment, as I wished it would, and so I smiled very weakly and gave halfhearted little waves to the folks who waved at me.

There were more forms to fill in, and, as I was doing that, pieces of paper were shoved in front of me for my autograph: a bizarre proceeding, signing for welfare while signing autographs. I had expected to be treated with deserved contempt, but I was amazed at the kindliness of these workers. Though dealing with families in the worst of conditions, and subject to abusive language, sometimes physical attacks, they seemed to maintain a good measure of compassion for the suffering humanity thronging their offices. Several of them assured me that they had seen other actors and well-known people who had come on hard times, but they had overcome them, and I would too.

That was about as cuckoo a morning as I've ever had in my life. In a few days, I was approved to receive three hundred dollars per month, plus food stamps totaling around thirty-seven dollars, enough to keep the *Canis lupus* from taking over the kitchen. Without projecting, or taking the Cassandra view of life, I was set to survive another day. The more I worked at AA's one-day-at-a-time philosophy, the better it got.

Taking a look at my past, something else we do there, didn't do much to bring balm to the soul, though. I was becoming aware of how detached I was from my children as they went out in the world and made lives of their own. Siobhan had gotten married, though I advised against it, and had made me a grandfather by giving birth to Fiona, a perky little baby, but then the marriage foundered. Malachy Jr. was off on his journey westward; Conor was studying filmmaking at NYU; and Cormac was still at Stuyvesant High School. They all moved and lived in their own circles, and it suited me not to intrude.

Of course, they knew that Diana and myself were not in the normal state of marriage and, of course, they didn't want to intrude on our troubles. I didn't tell them, or anybody else, that I was a welfare case.

One day, I got a call from my friend Mike Gibbons, an Estée Lauder executive, to inquire if I'd be interested in playing in a celebrity croquet tournament, a big society do. He explained that on a certain day every year, the croquet club invited celebrities to team up with the members, and fun was had by all on the lawn in Central Park. I was very flattered to be considered a celebrity, and promptly said yes.

Mike told me the dress for the day was white tennis shoes, white pants, white shirt, and, if necessary, a white cap or hat. It was to be that Sunday, and, as I didn't have the needed uniform, I made a swift trip to the Salvation Army clothing department to get the cut-rate stuff.

The only experience I'd had with croquet was one game at the house of some friends, Paul and Sasha Mayer, in the country. I had no idea of the emotions that storm in people's beings when they step onto a croquet court. To me, it had always seemed a la-di-da sort of activity, a trifle on the sissyish side, but when I saw the intensity with which Paul Mayer approached this la-di-da game, I realized 'twasn't a game at all. It was war—all-out, non-bloody war, where everything went and was accepted as fair.

I mentioned to Gibbons that my experience of this game was extremely limited, but he assured me that the members were aware of the shortcomings of the visitors, and that a skilled member of the club would look after me and cover up my mistakes. "Right you are," sez I.

On the appointed Sunday, I donned the Salvation Army slightly off-white garb and headed to Central Park to mix with society and celebrities. There were refreshments, of course: tea, coffee, sandwiches, wines, champagnes, and beers. Very tempting, too, but I stuck to the tea.

I was assigned as partner to a very charming, compact woman named Arrigan, who was also secretary of the club, and a skilled, aggressive player of the game.

One of the members, from Florida, a well-dressed lout, was loudly lamenting the admittance of unskilled people into the club and, furthermore, there were nouveau riche types getting in, people with no social background, and he didn't know what things were coming to.

Because of the large field of competitors, they had to change a rule to allow a shortening of the game, which caused the loud yahoo to bluster some more, as if the world was coming to an end. All I knew was, you got a mallet and were assigned a colored ball, and you whacked the ball through a series of hoops stuck in the grass. If you can, you also whack the opponent's balls as far as possible.

Mrs. Arrigan was terrific, and got us off to a good start, so we won, and we won again, and finally went up against the loud society yahoo. It was a cinch that he and his partner were going to win, but, as the last ball was going to be put through the hoop, I had one last chance. With all the energy I could summon, as far away from the final hoop as I was, I gave the ball a tremendous whack and knocked the loud one's ball out of contention, and in went mine through the hoop to win.

There was a gasp. The yahoo went purple in the face and said, "Rules should not be changed," etc.

The CEO of Rolex was in from Switzerland, and he was presenting the prizes. I assumed I was about to receive the first Rolex I ever owned, and maybe I could put it in the pawnshop. When my name was called, I stepped up to receive my prize, and was astonished to find myself being handed a magnum of Dom Pérignon champagne.

"What am I supposed to do with this?" sez I to the CEO of Rolex.

"By all means, drink it!" he laughed, and clapped me on the back.

"Yes, indeed," sez I to myself.

The real prize, of course, was knowing that if the yahoo from Florida even suspected that one of the winners was on welfare, the seat of his perfect white trousers would require a prompt and thorough cleaning.

There was a mention of me in one of the Florida society pages and, to the merriment of some of my friends, my picture appeared in *Town & Country*, a glossy magazine devoted to the doings of the rich and anonymous.

I was having some communications with Diana, neither friendly nor unfriendly, more businesslike than anything else. Though I couldn't dispute her contention that I had broken my word to her, and that I had serious problems with money, simmering in my subconscious was the conviction that, wrong as I was, it was no grounds for breaking up a marriage. Of course, I didn't think there was any reason for breaking up my first marriage, either.

One day, we met for lunch, and, looking at this beautiful woman who had spent twenty years with me and given birth to our two sons, I was again flooded with rampaging emotions, love for all that she was, rage at myself for losing her, anger at her for leaving me, and sorrowful nostalgia for all we had shared in years past. Nostalgia is dangerous when the present has to be dealt with, and an easy out for the wrongdoer.

Diana was in no mood for sentimentality, and wanted to deal with practical matters of the moment. In my most grandiose fashion, I announced I was moving back into our apartment, as I was weary of living on other people's charity, and there was enough room for us to live separately in the one apartment. An absurd idea, angrily delivered, which caused the object of my acrimonious affections to pack up her belongings in the middle of lunch, leaving me feeling pretty foolish.

Sobriety came to my aid, and I was later able to sit down and write her a letter, apologizing for my rotten behavior and assuring her that I had no intention of invading the apartment under any circumstances. I received a very warm letter in response, thanking me for writing and for all I'd done for Nina over the years, and essentially leaving the door open for future communications.

When my first marriage was shattered by my lunatic antics, I thought I could repair it with bravado and more lunacy, which landed me in the celebrated Tombs prison on several occasions, so I wasn't going to do that again. The alcohol which had fueled that trajectory was now out of my life. For the first time in my life, I began really praying for guidance to a God in whom I still had very shaky faith.

Going on the principle that it's never too late to change, I decided to court her as if we had just met. I would, as best I could, change the old behaviors. So began a courtship where no promises were made, only changes in my life. And we talked and talked, and, very gradually, a friendship developed, and a new kind of affection crept in, and then, what I now know to be love came to embrace me.

My courtship of Diana was proceeding quite well, and I was careful not to pressure her into any immediate decisions. It was a whole new beginning for me, because I was able to let real love show through. I had avoided it, I came to realize, because I'd learned that whomever you love will inevitably leave you, and whatever you love will be taken away, so it was best not to get too attached.

I came to appreciate more what Diana had been doing over the years, while I was gadding about, both while drunk and while sober. She had raised our children and took care of me when I needed it; she looked after Nina, and ran her business, Dovetail, which, though that was the year she and her partner shut it down, was at times our sole support. I came to realize that all she brought to my life wasn't my due, but a privilege, and one I'd sometimes done a poor job of earning.

I began to change my behavior in ways that weren't possible for me before. In the past, my reply to every criticism had been, "You knew what I was when you got involved with me," the spousal equivalent of the old parental saw, "You made your bed, so now lie in it." But that self-righteousness began to fade as our courtship proceeded. For once in my life, I tried to listen and, not alone that, I tried to hear what this woman had to say, and that's a feat for a world-champion talker and a self-righteous Irishman.

It took a while, but eventually Diana invited me to move

back into the home where we had lived for over twenty years. Of course, I moved with alacrity on the invitation, with great gratitude at being offered the second chance. But no sooner was I settled in with Diana and feeling happy with the ensuing domestic harmony, than a series of jobs took me away from home again.

First, a call came from Chicago for Frank and myself to perform in *A Couple of Blaguards* for our old pals Sheila Heneghan, Michael Cullen, and Howard Platt. It was a delicate decision, considering the very recent reconciliation, but Diana was amenable, so I went. It was quite a successful run, and not alone because it was extended from a few weeks to a few months. Diana came to visit me there, and it seemed as if the slow pace had been the best route toward reconciliation, as we began to grow closer than we had ever been before.

Then, I was called off to Ireland, where along with making a fine film, I learned that it was not just my feelings about people that had needed to change, but my feelings about places, too.

Jim Sheridan was to direct a movie called *The Field*, with a splendid actor named Ray McAnally. Sheridan told me there were a few parts I'd be suitable for, and I mentioned to Richard Harris that I'd probably be in the film. He said that the McAnally part of Bull McCabe was one he'd like to do if anything ever happened to Ray.

When I next rang Sheridan in Dublin to inquire as to when he would need me, he could hardly speak. Ray McAnally had died. Good citizen that he was, the last thing Ray had done was vote in the Irish national election. There were years when he had had to fight his demons, but he was in the end a sober man, happily married, who loved a bit of gardening. Artistic respect was coming his way, and he'd more than held his own in *My Left Foot* opposite Daniel Day-Lewis, and now this tour de force had been offered him, but it was not to be. Life is not fair, and neither is death, for it struck down this grand man at a comparatively young age, just as he was about to enter a new phase in his life.

That was the news Sheridan had for me—bad news for everyone. A huge crew had been employed and a cast engaged, locations scouted and rented, and all set to go, but the death of one man stopped all of it. I didn't mean any disrespect to Ray McAnally, but I mentioned to Jim what Harris had said about doing the part if anything happened to McAnally. Jim demurred by saying that he was not sure if Harris had the capabilities to do this demanding part, and we left it at that.

My next call was to Harris to accuse him of putting a curse on another actor just to get a part. He didn't see the humor in that, nor did he at first believe me when I told him the news. But before we got off the phone, he said he would do something he hadn't done in thirty years—audition. He got on a plane to London and informed Jim Sheridan and Noel Pearson, producer of the movie, that he was ready to read for them.

Pearson said "Fine!" but Sheridan was reluctant. They popped over to London and met with Harris and auditioned him. As he was about the only actor available, they hired him, and so goes show business. Aside from chomping on a herd of cattle and some odd bits of scenery, Harris did an excellent job.

I met with Noel Pearson in New York City to discuss my participation, and was told there were two parts I could choose from, a police sergeant and a pub keeper/auctioneer. I opted for the pub keeper/auctioneer, but my good friend Harris pushed for an actor he'd seen at the Abbey Theatre one night, and so I had to settle for the smaller cop role. Fair enough. I was delighted to work for Pearson, and with Sheridan again, and to be doing my first film in Ireland.

It was a glorious couple of months for me. I worked not more than one or two days a week, and so got to explore spectacularly beautiful parts of Ireland, landscapes from which the dark and bitter clouds of memory were finally lifting. During that protracted visit to Ireland, not once did I take Guinness, the wine of Ireland, into the gullet, nor did I put the whiskey into the system, not a small victory. I left the country more peaceful and serene than I had been for a long time. No longer was I trying to change the past. A friend, Lynn Fine, said, "You know you are maturing when you realize you cannot manufacture a better childhood."

With my newfound benignity toward Ireland coursing through the system, it was easy for me to return there for the premiere of *The Field*, a very glamorous event in Dublin. I was even able to go back to Limerick without the rage and rancor poisoning my bloodstream, as had always happened before. There were always kindly people there, but I hadn't been able until then to begin to appreciate it.

My next journey was to San Francisco, where my brother Mike was going to produce *Blaguards* with a friend of his. With the usual trepidation and great faith, I set out in a used car I'd bought for the cross-country jaunt. Stopping to get gas and have the oil checked during the early portion of the journey, the service station attendant told me there was a mouse hiding under the hood. So, on the way, I fed him whatever I could poke into the little hole in the hood, and I soaked tissues in water so he wouldn't dehydrate. How he survived the heat was a mystery to me.

On the way, I stopped for AA meetings whenever I could. I recited the Serenity Prayer in meetings all the way from New York to California. "God grant me the serenity to accept the things I cannot change, the courage to change the things I can, and the wisdom to know the difference."

Having been to a particularly good meeting in Cheyenne, Wyoming, I decided to keep driving that night, up into the mountains and as far as possible 'til the tiredness got ahold of me. It was a lovely western night, an Indian's dream, the sky aglitter with stars, and a good-size moon looking on benignly. The car was smoothly purring up the mountain, when there was a loud bang and I had to use all my strength to keep control, as the vehicle shuddered and skidded, and the metal rim of the wheel screeched against the road.

When I managed to stop the car, I got out to survey the damage and found the right rear tire in tatters. I made an attempt to change it, but the jack didn't work. Of course, I envisioned myself spending the rest of my life stranded in the Rocky Mountains. (At times, it's best to think the worst, so then what's actually happening won't seem so bad.)

Standing on the steep incline of that mountain at 1:00 A.M. with a broken jack didn't allow me much hope, but stars were near enough to touch, and I imagined the higher power couldn't be too far away, so I began bellowing the Serenity Prayer. As I raised my voice, I could hear it echoing off the cliffs and jumping back from the valleys, booming from in front of me and behind me.

"God grant me the serenity to accept the things I cannot change, because I cannot change this fucking tire." 'Twas then I heard God laughing wildly and boisterously, and I found myself joining in.

Looking down the steep mountain, I then saw the flashing lights of some official vehicle, what turned out to be a state trooper. He pulled over and stepped out of the car, almost a stereotype of the western galoot: tall, rangy, weather-beaten face, cowboy boots, and Stetson.

"Are you all right, sir?" sez he.

"I had a flat tire, and . . ." sez I.

"The question, sir, is, are you all right?"

I said I was fine. "Good," he sez, and then he asked what the problem was with the car.

"A flat tire," sez I, "and I can't work the jack."

No problem, he tells me, and he gets a big jack out from his trunk, pumps my car off the ground, and in double-quick time changes my tire.

I told him he was the tallest angel I'd ever met, and asked whether he would be offended if I offered him money. He replied he wouldn't be offended, but he couldn't accept any. If I said a prayer for him, he told me, he'd be sufficiently rewarded. It was a small payment for a big service.

Subsequently, I wrote to the governor of Wyoming to praise this man who had gone out of his way to be of service. At other times in my life, the sight of flashing lights and tall policemen would have evoked a different response, but sobriety changes circumstances and perceptions.

Somewhere in Utah, my rodent traveling companion disappeared, leaving me wondering how a small mouse from Maryland was going to adjust to the salt flats and Mormonism.

I traveled the rest of the way alone, and settled into San Francisco for a couple of months, getting to visit with the brother Michael and his wife, Joan, and their children, Mary, Angela, Michael Jr., and Katie. The brother is well-known for his sharp, sometimes acerbic wit, and many took it as a great compliment to be insulted by him. When the folks at the San Francisco newspapers were caught short for a quote, it was not unknown for them to go to Mike to get a pithy observation. The dean of gossip columnists, Herb Caen, was often seen at the bar where Mike worked, listening to glean a bit of the wisdom.

As it was for me in Ireland, being in San Francisco and staying sober was a minor triumph. All previous visits involved drinking, and stupid, disagreeable scenes with Michael over issues that neither of us could remember the next day, though enough would be recalled to leave a nasty resentment. It's remarkable how the brother Michael became so much more reasonable when I stopped drinking.

Diana was able to join me again for more conjugal visits, and from there, I made my way home.

In New York, I was cast in another play, called *Remembrance*, which was produced by Nye Heron at the Irish Arts Centre. *Remembrance* was written by a nasty little man named Graham Reid, another North of Ireland Protestant. In the play, an elderly Catholic woman, played by Aideen O'Kelly, goes to a cemetery every week to visit the grave of her son, who had been killed in sectarian violence. I played an elderly Protestant man, who visits his son's grave at the same times. They meet and fall in love, against all the sectarian prohibitions of that part of the world.

The production was a huge success. We garnered great reviews and had a good, long run, but when the playwright, Graham Reid, came over from Belfast to see it, he roundly attacked us for doing it the wrong way. He criticized the actors individually, and found something wrong with every one of us. He told me I was too "Catholic" for the part.

At first I thought he was twitting us, but then I realized the little prick was serious! He had written a play that was a plea for tolerance, understanding, and love, and here he was, your average nasty bigot.

Of course, I proceeded to share with him some of my key thoughts as to his nature, and when the time came to move the production to a bigger theatre, I was not among those considered to play my part. They also insulted Aideen O'Kelly by asking her to audition for a part she had played for nine months. And people think this is a glamorous profession!

Shortly thereafter, I went to Los Angeles again, at the behest of my managers, who were moving their main operations to the West Coast. It was a highly speculative move, as the prospects were not golden, and once more it involved a separation from Diana, who had obligations of a family nature. Her mother was getting along in years and, though the boys were grown, there was always the task of making sure that Nina was being looked after in her group home, so I had to go it alone.

I'm not sure which occupation has the top rate of unsuccessful marriages, but I'm sure that if the entertainment business is not the leader, it does not lag far behind. At any given moment, there are about two hundred thousand people filled with dreams, expectations, and hope that the spotlight is about to come around and shine on them and lift them to stardom. It happens to some, but for most, when the spotlight hits, there are sharpshooters perched in the watchtowers, waiting to mow them down.

Still, we go on and make forays to California and returns to New York and runarounds to casting calls, 'cos you never know, you never know.

But sometimes, you eventually learn something. During these trips, I finally made up my mind that the Hollywood scenario was long past for me. My life, my love, and my destiny are bound up in New York, because that's where Diana, the woman of my life, resides, as do most of my children and so many of the friends I've made (and I would like to have a good turnout at my funeral).

I found I was able to scoop out a reasonable living at the trade of my choice. Bits and pieces of soap operas came my way, plus a week here and there on films, and jobs for various stage shows. Most important, though, was that I continued to build and rebuild the trust that goes into a marriage.

Sobriety had taught me to watch out for fantasy and unrealistic expectations, as an expectation can be a resentment in the making. Bearing that in mind, things were going well in my life, in my work, and in my relationship with Diana. But there are obstacles unforeseen, and, of course, unplanned, on the road to redemption.

I thought I was thoroughly and absolutely familiar with depression. Hadn't I felt it once in a while, and isn't depression an inevitable part of the hangover, and hadn't various people over the years talked to me about the condition? My own daughter, Siobhan, had called me a few times from Boston, trying to enlist my help in recovering from her own dark tale, but I wasn't listening and just gave her some swift and thoughtless advice about snapping out of it. I was to learn just how thoughtless those words were.

It happened on the twenty-third of July 1994, a very hot Saturday in New York. I had slept late, and Diana was off visiting her mother in Pennsylvania. The day before had been pleasant and satisfying, as I'd celebrated nine full years of abstinence from alcohol and felt I was well on the road to sobriety. It had been three thousand, two hundred, and eighty-six days without a drink, give or take a leap year, and every day a day to be treasured.

I'd made some tea, had the egg or two, read *The New York Times*, and was feeling quite at peace with myself. It was getting pretty hot, so I proceeded to the bedroom, turned on the air conditioner, and began tidying up, when suddenly—and very suddenly it was—I felt as if I'd received some sort of blow to the solar plexus. It wasn't physical, but it doubled me over as if it were, and the whole world seemed to go a grim and ghastly gray.

I sat down on the side of the bed, whimpering like a wounded animal, rocking back and forth in fear and panic. Inside my chest, I became aware of a vast, dark emptiness, in which my heart was being grasped by a huge, cold, steel fist, and slowly squeezed. I was aware of no physical pain, had no sense that I was having a cardiac seizure, a stroke, or anything like that; there was only a vast psychic panorama of bleakness, an icy tundra of grayness, and whited skeletons hanging from dead trees, condemned to eternal despair.

I crawled into the bed and covered up my entire body, and shivered with the cold and perspired with the heat simultaneously, as if I had malaria or some strange tropical disease. There was no doubt in my mind that I was plunging into depths from which there was no return. I knew I would never recover from what had just happened, and I wondered how I could not have known I had been perched on the edge of a precipice overlooking hell.

I had been seeing a doctor in New Jersey, a Dr. Joel Fuhrman, and I was doing well under his ministrations. As well as being a sober, recovering alcoholic, I was now a nonsmoker and a vegetarian who had also stopped drinking coffee. I was exercising, and the pounds were dropping from the overweight body. In every way, for the first time in my life, I was engaged in a focused and concerted struggle to do the best I could, so why had darkness descended on my life at this time?

I knew there was no cure for what had just hit me. Friends had told me of their precipitous tumbles into the darkness, and how they had climbed back, through the grace of God, sobriety, and modern medicine, but I knew there was no point in calling them. I knew my case was hopeless, beyond the help of God, and certainly there was no medicine in the world that could cure me.

It is at vulnerable times like these that the other disease, the disease of alcoholism, likes to come out and play. The alcoholism says, "Nothing like that ever happened to you while you were drinking." Indeed, it seemed the best thing I could do was get out of the house, go to the saloon, and drink until I felt the weight of this darkness no more. Those thoughts came dancing through the mind on that strange day, and yet even the promise of a drink was not enough to get me out of my bed and out into the world.

At one point, I looked at the clock and saw that it was three o'clock, and it was totally dark outside. When the depression hit me, it had been two P.M. on a Saturday afternoon—so many hours had fled! I tried to sleep, but the fear of facing another day put the iron grip on my heart, and so went the rest of the night. Something informed me I would never sleep again.

When Diana returned that Sunday evening, I behaved as normally as possible because I was not alone depressed, but also ashamed of the way I was feeling. The normalcy act didn't work for long. I'm blessed in being married to Diana, whose intuition can sense a mood change from three thousand miles away, and who doesn't brook any nonsense about my being "all right" when it's not so.

I found myself babbling to her, embarrassed that not even she could cure or dispel this thick gloom. She got on the phone to Dr. Fuhrman, told him the situation, and made an appointment. I didn't see much point in visiting the doctor, as it seemed to me that for all intents and purposes, my life was over. There was hardly anything I didn't worry about. For instance, I knew that I would never work again, and we wouldn't be able to afford the rent, and where in God's name were we going to store the books after we were evicted, not to mention the furniture and beds and our other belongings? For the first time, I contemplated suicide in a serious way. A parody of the familiar sonnet took over the head: "How do I kill me, let me count the ways . . ."

I did manage to see the doctor, and he prescribed a new anti-depressant called Serzone, reputed to have hardly any side effects. Of course, I wanted simply to pop a pill or two, expecting that the sun would shine into my soul again in double-quick time, but it didn't work that way. The good doctor Fuhrman said it could take from six to twelve weeks for the pills to work, a bit of news that left me groaning, as I didn't see how I could keep on living that long.

To cap it all, my memory, hitherto good and capable, began to fail me, and going for auditions was an agonizing and painful experience. I'd had a reputation for having a good sense of humor, but that disappeared as well, as I couldn't find humor in anything. Some folks were doing a satirical film based on some right-wing doings in the U.S.A., and I went to read for the part based on the rabid Reverend Charles Coughlin. The humor was lost on me, and the producers wondered to my manager why a humorless, heavy-handed dolt like myself was even sent for an audition.

In the newspapers, I saw nothing but accounts full of dreadful doings portending the end of the world. Books no longer interested me, as I couldn't hold the characters and events in my mind for even a few minutes.

Diana insisted we take a trip to Maine, where we stayed in a very rustic cabin, close to the woods and close by the sea, bucolic and beautiful in every way. There were whales to be seen and dolphins leaping gracefully, but I couldn't keep my eyes focused long enough to wonder at the grandeur all around me. My neck seemed too weak to hold my heavy head, and I walked around those sunny days with my head hanging down, as if I were deeply shamed. Diana tried and tried, with love and patience and caring, to bring light back into my life, to no avail.

Back in New York, the friends and family rallied around, but the suffering was relentless. Every morning at three A.M., I awakened and began my Beckettian inventory of disasters that would arrive with the coming day. The list of the disasters to come was endless. My son Conor had joined the New York Police Department, and I knew that something awful was going to happen to him. Cormac would be unable to support himself, and Malachy Jr. was off diving at the Great Barrier Reef in Australia, so I knew a great white shark would get him. Siobhan in Boston would never find a good career, and Diana would leave me, and my heart would give out, and perhaps I'd go blind and not be able to read, not that I was interested in reading. The goddamn antidepressants were not working, even though the dosage had been increased to the maximum.

But then the diligent Diana took over. She set out to do some research on what had happened to cause my catastrophic plunge. On a prescribed diet of vegetables and fruits, with no meats, fish, dairy, or oil, my cholesterol had dropped over a hundred points. Diana discovered that a sudden deprivation of cholesterol sends the liver into some kind of rebellion, causing neurons to wither away, leaving the body in confusion. She came to me with the immediate suggestion that I get some oil back into the bod, and that I load up on fish protein, which I did, and there began the sight of a flickering light at the end of that very dark tunnel.

It took about ten months for me to feel normal and get restful sleep again, but with the love and encouragement of Diana, along with the family and friends, I've left those dark, despairing days behind me. Well it was, too, because the ride never gets any less bumpy, and all you can do is learn to be grateful for the ride, bumps and all.

Age and knowledge do not necessarily imbue us with wisdom. I've stopped asking the reason for my existence here on earth, and where I came from, and why and where I am going.

I know now that I became an actor because I didn't want to be Malachy McCourt, and I became an alcoholic because I didn't want to be Malachy McCourt. But, as they say, no matter where you go, there you are, and therein is the solution to most of the problems in my life: acceptance of myself as I am this day, and gratitude to whoever or whatever put me here.

There was one Christmas when I was, as usual, feeling pretty miserable, so I sent out a handmade Christmas card which said, "Despair is our only hope." The irony of it made people smile, but it was too close to the truth for me. These days, I don't think that way. To me, despair has become a sin.

It's taken a lifetime to unlearn the lessons and break the habits that were handed down to me growing up in the lanes of Limerick. It has been pointed out to me more than once that I trod the same path as my father. All of us are scarred, but some go to extremes in the quest for the happier childhood they never had.

I have learned that we're not born sinners who deserve our suffering, and I look to see where I can bring comfort to others. There was no touching in my family when I was growing up; there was no hugging of children, no kissing of parents. When I was a child in Limerick and told my mother I loved her, her response taught me that that was not a thing to tell anyone else. Now I try to teach the small ones just the opposite.

My son Cormac has made a place for himself in New York as a massage therapist, healer, and yoga master. His daughter, Adrianna, a beautiful, sunny girl, is now three, and often visits with Diana and me. She's right at home sitting in my lap, as I read a book to her, and she's happy to give her grandfather a kiss on the cheek. I'm even happier to receive it. By showing Adrianna the love she deserves, I do my bit not to pass on to her the inheritance that sent me careering about the world in a self-destructive, self-deluding, alcoholic fog.

It is a great blessing to have arrived where I can do that for her, and it's good I did it sooner rather than later. When I'd just gotten out of the hospital, there was no crawling up into the lap for the granddaughter; indeed, for a time, there was some question as to whether there would be any lap to sit in at all.

The latter part of the nineties were very significant for the McCourt clan, as if that would be news to anyone by now. It started one day when Mary Breasted, a fine novelist and a former reporter for *The New York Times*, asked the brother Frank what he was up to now that he had retired from the teaching profession. He was happily remarried then, to the lovely, bright Ellen, and as he had the time, while she was off working for public television, he was engaged in writing a memoir. Mary, being aware, as many were, of Frank's way with the language, asked if perhaps he had a hundred pages or so to show to an agent. The man said he had, and bunged them off to her. She, in turn, took them to an agent, who said the whole thing was "too Irish," and there would be no market for such a book.

Mary, not to be deterred, persisted, and *Angela's Ashes* was published. All of a sudden, this retired schoolteacher was in demand to speak at venues from New York to Helsinki, and he was showered with awards.

As his fame and recognition grew, the rest of the family were caught up in the reflected glow. My son Conor, who had not forgotten what he'd learned at NYU film school before joining the NYPD, had been working on a documentary about the family, *The McCourts of Limerick*, and now he completed it and sold it to HBO.

A friend, Charlie DeFanti, approached me and asked me about writing a book; a friend of his, John Weber, ran a small trade press, and would be delighted to publish my efforts. I told him that talking was my area of expertise, not writing, but I'd have a go at it, and I did.

After I'd written a bit, an agent, David Chalfant, came aboard, and it was his opinion that this book needed a major publisher. Arrangements were made, and after a bit of bidding it was published by Hyperion, and that's how *A Monk Swimming* got on the bookshelves and hit the best-seller lists.

But life is never unalloyed joy. Before setting off on what was to be a triumphant book tour, I took the precaution of having the medical checkup. The physician, a courtly gentleman, went about his task of examining the bod with care and thoroughness. There is not a man of my acquaintance who is exuberant at the prospect of a prostate exam, and I'm no exception in that department. Besides, there seemed no reason for me to get prostate cancer—that sort of thing was for people who neglected themselves.

However, the good doctor informed me that there was a spot of roughness, and the blood test hadn't gone well—not good, but this still wasn't a 100 percent indication of cancer. So, I went to see a specialist, who took a biopsy to make a complete diagnosis. During all of these exams, I was pushing away the thought of having cancer of any sort. I said to myself, "Not now, and, indeed, not ever," and that was that.

My biopsy was performed on June 16, 1998, which, as you know, is the anniversary of Bloomsday, the very day Leopold Bloom and Stephen Dedalus wander about Dublin in *Ulysses*. Immediately thereafter, I went off on my book tour, determined not to get down or discouraged by the prospect of prostate cancer.

I went to Boston, New Orleans, Dallas, Denver, Chicago, St. Louis, Seattle, and Washington, D.C. I read to large, enthusiastic crowds, answered questions, and patiently listened as I was told by hundreds of people how much they loved my brother's book. That's when I got the idea of telling people my next book would be called *I Read Your Brother's Book*.

A fair number of critics and some of the public blasted me for not writing the same kind of book my brother wrote, and for not having his talent. Indeed, I was blamed for not being my brother. I now pledge to all those naysayers that someday I will write *Angela's Ashes* and change my name to Frank McCourt, 'cos I'm devoted to making people happy.

I was having a grand time signing books and meeting the people of America, talking, singing, chatting, and shaking hands, keeping the specter of cancer from impinging on a once-in-a-lifetime traveling party, but it was not to last. Prostate cancer in an older man is like having mice—if you think you have them, you do. The doctor instructed Diana to inform me that I did indeed have the malignancy on all sides of the prostate.

Diana was most comforting and caring on the phone as she

broke the news to me, and that was a bit of a buffer against the shocking reality. But no matter how I moved the words around or rearranged them, they metamorphosed back to "You have cancer." Many people are told that same thing every year, but it's no comfort to know that at first.

I was in St. Louis, Missouri, when I got the news, and a minor glitch in the hotel arrangements took my mind off the trouble. Someone had canceled my hotel reservation, and there was no other room available. I went into a shouting rage on the voicemail of my publicist in New York City (for which I later apologized).

Now, I have slept on the ground, under overhanging rocks, in tents, and in caves, so a canceled hotel reservation is not a major calamity for me, but it allowed me to vent and howl out my disappointment. Here I had thousands of days of sobriety to my credit and I was nicotine-free. Fidelity in my marriage had become vital to my spiritual growth, and I was either making amends or attempting to make amends to those whom I had injured or hurt. I was on the crest of a wave of affection for the McCourt family, and hadn't I even written a best-seller myself?

It wasn't right! I felt I'd had enough fucking suffering in my life, and I just wanted a chance to enjoy myself and to put things right.

I railed within at the injustice of it all, as I had already decided that my cancer was incurable, and that there wouldn't be many days left. All this stuff was avalanching through my mind, as I sat in the lobby of the hotel, waiting for them to find a room for me. They finally did, and I sat down at the desk to evaluate the whole situation.

The first thing to do when I got back to New York was to make the will. The next order of business was my funeral, and a list of people who would speak and sing. Then, I had to make a list of people to whom I still owed amends. I got so engrossed in the prospect of dying that I lost sight of the fact that people get cured and recover from cancer, as they do from other diseases.

After finishing the book tour, I informed friends of my condition, one of whom, Maureen O'Brien, my editor at Hyperion, offered me Michael Korda's book *Man to Man*, about his experience with prostate cancer. Despite the millions who get cancer, there is always a sneaking suspicion that you are the only one who knows what it is all about. The emotions you experience are the opposite of those you feel the day you become a father for the first time, but the feeling of uniqueness is similar. As I read Korda's book, that feeling dispersed.

I got on the blower to him at Simon and Schuster, and, to my astonishment, he answered my call. He recommended that I see his physician, Dr. Russo, and with Diana by my side, as she would be at all the meetings and conferences with doctors that were to follow, I learned the various treatments for prostate cancer, which depended in part on where the little cells are residing. The options ran from surgery, where the whole prostate is removed, through radiation therapy of various sorts, to watchful waiting, which is simply doing nothing, as many men die with prostate cancer but not that many from it.

The compassionate Dr. Russo said that, as much as he loved the surgical procedure, he would not recommend it for me. Diana and I thrashed out all the possibilities, including dying during the surgery, which does happen. I ultimately opted for the implantation of tiny radioactive seeds, as the least damage is done with that course of action.

When the day came, I taxied off to the hospital, holding hands with Diana.

I can tell you that it's not a bit dignified having people fixing the prostate gland. Entire phalanxes of masked medicos learn what parts of you look like that you've never seen yourself. Somebody tethers you to the monitors, and another masked bod, brandishing a hypodermic needle, tells you that it won't hurt a bit. It didn't, and whilst chatting away, I suddenly went into a warm darkness and snuggled into the deepest sleep of my whole existence.

I was awakened by a voice singing, "Will ye go, Lassie?"—not too well I might add—and discovered that the voice was mine own. (It would not give Pavarotti a single moment of unease.) I floated 'twixt consciousness and its opposite for the next three hours. When it was time for me to go home, the good doctor arrived to give instructions for the specific short term and for the general long term, then he told me to go off, lose weight, and enjoy the rest of my life. After our talk, there was hardly any doubt in my noggin that I was well on the way to recovery, and, outside of a few passing inconveniences, there's been no discomfort or negative consequences from the procedure.

Thinking you're about to take leave of it is always a fine motivation for a look back at your life, and that look led me to conclude that alcohol was a major cause of many of my troubles. As an alcoholic, I was a liar, a cheat, a swindler, a thief, a deceiver, and just about anything else that a person can become when his defects are allowed to run wild. But getting sober did not mean I suddenly became a haloed saint, or that all the defects took a hike into the wilderness, never to be seen again. Not bloody likely! You can take the brandy out of a fruitcake, but you still have a fruitcake.

Getting sober has allowed me to work on those defects and,

bit by bit, to get better. I've learned to keep an eye out for all the things that people do that are right and good, and act accordingly. It's fairly clear to me now that if I stay clean and sober, I've got a good chance of acting right, and having a peaceful day, and then tomorrow, if there is one, I won't have to lie about what I did the day before.

Thoreau said, "Simplify, simplify, simplify," and I've simplified all my needs and wants to love. The love of my wife, Diana; of my children, Siobhan, Malachy, Nina, Conor, and Cormac; of my grandchildren, Fiona, Mark, and Adrianna; of my brothers, Frank, Michael, and Alphonsus, and their families; and of all my friends.

A Benedictine monk named Father Basil told me once that there's another life after this, and that when I reach it, the only question the deity will ask me is, "Did you have a good time?"

While I'm in no hurry to get there, I'm ready to answer, "Yes, I did, thank you very much. I particularly liked those parts toward the end."

Acknowledgments

Many thanks to Howard Mittelmark, my guide and collaborator on the journey through the words. To the Sandbergs—Betty, Carl, Carin, and Jackson—for opening your door to me. To Heidi and Kelly for your generosity of spirit. To Marcy and John Carsey, my loving friends of many years. To Bob Schilling, for the years of friendship. To Mauro DiPreta, editor supreme, a grand man of patience and skill. To Paul Mayer, and the wonderful Sasha, who died on us. To David Chalfant, the intrepid agent who did it again. Dr. Harvey Benovitz who found it, Dr. Michael Zelefsky who cured it. The grand cast of *Remembrance*: Ann Dowd; John Finn; Terry Donnelly; Ellen Tobie; Aideen O'Kelly; our director, Terry Lamude; Brian. Sheila Brown for leading the cheers. Isaiah Sheffer for making me a star at least once a year at Symphony Space. Peter Sheridan, the man who makes Dublin your home. Mickey Kelly, fellow Blaguard and friend. Bob Miller for the pen that wrote this book. To Terry Moran, Gene Secunda,

Acknowledgments

Dennis Smith, and punster Pat Mulligan, and all the First Fridayans. Susan Cheever, my pal. To Charlotte and Ciaran at the Irish Repertory Theatre. To Pauline, Neal, and the crew at the Irish Arts Centre. To Charlie DeFanti, friend, mentor, and guide: What would I do without you? To John Weber, whose generosity is unmatched, for the continuing faith. To Jason Miller, Bob Schlesinger, and Agnes Cumming, and all the talented members of the Scranton Public Theatre, who made *Inherit the Wind* a glorious experience. There is no place on earth like the world. . . .

To the lads in that other club: Jack Z., Duane, Rudy, Michael G., Brian M., Chuck, Kevin, David T., Andrew K., Peter B., Peter Y., Bruce B., Glenn, Anthony G., Michael M., Sean M., Pat H., Joe S., John H., Terry L., Charlie M., Charlie H., Ron F., Ronnie, Jack W., Rick, Eliot, Danny, Joe C., Jim D.—you all know who you are. To my friend the late Mike Molloy and, indeed, all the wonderful Molloys. And to my brothers, Frank, Mike, and Alphie, and the women who make them happy, Ellen, Joan, and Lynn. And always, to Diana, who makes me happy.